Conscience

ALSO BY PATRICIA S. CHURCHLAND

Touching a Nerve: Our Brains, Our Selves

Braintrust: What Neuroscience Tells Us about Morality

The Computational Brain (coauthor)

Neurophilosophy: Toward a Unified Science of the Mind-Brain

Conscience

THE ORIGINS OF MORAL INTUITION

Patricia S. Churchland

W. W. NORTON & COMPANY
Independent Publishers Since 1923
New York | London

Printed in the United States of America
First Edition

Manufacturing by Lake Book Manufacturing
Book design by JAMdesign
Production manager: Julia Druskin

ISBN 978-1-324-00089-1

W. W. Norton & Company, Inc., 500 Fifth Avenue, New York, N.Y. 10110
www.wwnorton.com

W. W. Norton & Company Ltd., 15 Carlisle Street, London W1D 3BS

1 2 3 4 5 6 7 8 9 0

Contents

Conscience

INTRODUCTION

Wired to Care

I cannot and will not recant anything, for to go against conscience is neither right nor safe. Here I stand, I can do no other, so help me God. Amen. MARTIN LUTHER

Hanyani Camp is a Déné village of about a dozen log houses tucked into the bush on the banks of the Nahanni River, in Arctic Canada.[1] I am waiting on the landing strip for a single-engine Beaver to fly me out to Fort Simpson. My wait could be longer than expected, since the Mounties have shown up with a prisoner and will take priority on any flight out.[2] Their prisoner is young, perhaps twenty-two years. He sits peacefully in handcuffs. He is mildly embarrassed, yet dignified and composed. The story? No one explains, or even hints with a bit of body language. I know enough, however, to make a reasonable guess. A brawl, probably; a murder, quite possibly. The two constables and the cuffs are a bad sign.

The man surely has a deep range of skills for living off the land. I can manage to put up a tent in the howling wind, but this Déné man, he can certainly read the river and the bush. He will have regularly brought in moose and tracked bear—skills I cannot begin to claim. He will know how to survive well through eight months of ruthless winter. I know precisely nothing of this Déné man or his circumstances. Still, I find myself spinning a hopeful narrative whereby he does not spend time in prison, closed in by walls, with no river, no moose, and no family.

My conscience roils as I recall the bitterly familiar story of how the life of the Déné people was shattered after "contact." Flourishing Déné villages were decimated by smallpox, crumbling the cohesion of traditional life along with deep survival knowledge and social wisdom. Déné hunting lands were "gifted" to the greedy invaders who plied the elders with whiskey and sneered at their skills. In the contemptuous name of "civilization," beloved Déné children were rounded up and placed in residential schools hundreds of miles yonder, where they were isolated from siblings and beaten for speaking their own language, the only language they knew.[3] I find myself remembering Jackson Beardy, a brilliant artist, snatched from his loving family in childhood. As an adult, he felt he belonged nowhere, estranged from his Cree family but never embraced by whites. In 1970, the National Gallery in Ottawa showcased his masterful paintings, but on opening day, the security guard refused him entry.[4]

I briefly consider shambling over to the Déné man, perhaps striking up a kindly conversation. Oh, get a grip on yourself. How presumptuous, condescending, and self-consoling. For one thing, he is apt to feel obliged to put out the effort to reassure *you*. The confused young constables would have to wonder if they should shoo this meddlesome woman back to her spot in the weeds. Despite my self-reproach, I feel that recognizable tightening in my gut. I ought to do something. But there is nothing to be done.

Were I a solitary creature like a salamander, none of this would trouble me. I would have no moral conflicts, no social conscience. I would feed and mate and lay my eggs. I would not fret about other salamanders, not even those hatching from my very own eggs. I would see to my own needs, and care not a whit for others. But I am a mammal, and like other mammals, I have a social brain. I am wired to care, especially about those I am attached to.

My mammalian brain sees to it that I am attached to my family and my friends. This means that I am concerned about how they are getting on. I have empathy and sympathy and sometimes moral indignation. I can be powerfully motivated to cooperate, even when doing so entails going against my own interests. My brain has also learned the traditions of my kith and kin. Consequently, I may be moved to tell the truth when a lie would be self-serving. I may be moved to punish those who torment the weaker ones or who exploit the gullible ones. I have a conscience. Or as I sometimes think of it, my brain sees to it that I have a conscience.

Some of the deepest thinking on human morality originates with the Greek philosophers in the fifth century BCE. These include Plato and Aristotle and, in his unique style, Socrates. Interestingly, the ancient Greeks did not have a single word that is equivalent to our word *conscience*. They did not need the word, however, to appreciate the persuasive influence of moral feelings. The specific word *conscience* was invented later, by the Romans. In Latin, *con* means "in common," and *scientia* means "knowledge."[5] So *con scientia* meant, roughly, "knowledge of the community standards." Like Socrates, however, the Roman philosophers knew that conscience cannot always conform to community standards, for our moral sense sometimes requires us to challenge those very standards.

Famously, Martin Luther (1483–1546), the priest and theologian who was a seminal figure in the Protestant Reformation, rebuffed prevailing church standards when he nailed his devastating objections to the cathedral door in 1517. The church's standards, especially as they involved moneygrubbing and craven obedience to authority, were precisely what Luther was convinced were morally heinous. Luther explicitly used *conscience* to mean a broader knowledge of what is morally right and wrong—a sensibility extending beyond prevailing standards. Invoking this construal of conscience raises the question of the source of human knowledge of right and wrong, if not community standards.

Socrates (469–399 BCE) had a lifelong concern with how we come by our moral convictions. He was especially troubled by our propensity to claim certainty about what is right and wrong, even when that certainty is misplaced. In revealing the pattern of his own moral deliberations, he referred to his *inner voice*, where we might speak instead of *conscience*. Always modest, Socrates explained that his inner voice was not entirely reliable and might sometimes lead him astray. Acknowledging the unreliability of his inner voice led Socrates to claim that moral wisdom necessitated admitting our moral ignorance and imperfection. Phony wisdom, he warned, had the look of dogmatic conviction. Certainty about one's moral stance might be soothing, but it tends to blinker us to damage we are about to cause.

Socrates did not imply that his inner voice speaks only about moral questions. One's inner voice can examine both sides of a moral issue, but it can also chatter on about a lot of things, both moral and practical, both sensible and foolish. On financial matters, my inner voice often sounds like my thrifty dad: "You change the oil yourself, dearie; no need to pay someone to do it for you." When I am writing prose, my inner voice sounds like Mrs. Lundy, my grammar teacher, catching the misplaced modifier. Quite often, the voice sounds just like me talking to myself: "Look on the funny side" or, during the years when philosophers bashed me for studying the brain, "Outlast the bastards."

Sometimes my conscience is not a voice at all, but an uneasiness, a nagging sense of something needing to be done, or maybe something needing to be avoided. Or it is the way an image keeps snagging my attention. Like a mental song that I cannot shut off. As Paul Strohm ruefully puts it, the "characteristic habit of conscience is to goad, prick, wheedle, denounce and harass, rather than to mollify or assuage."[6]

Is there a precise definition of the word *conscience*? As with many words, such as *vegetable* and *friend*, we know well enough what it means, despite lacking a precise definition. Because the meaning of *conscience* may vary in different ways across cultures and subcultures, as well as across time, for present purposes I favor adopting this working formulation: conscience is an individual's judgment about what is morally right or wrong, typically, but not always, reflecting some standard of a group to which the individual feels attached. The verdict of conscience is not solely cognitive, moreover, but has two interdependent elements: *feelings* that urge us in a general direction, and *judgment* that shapes the urge into a specific action.

For children, learning the word *conscience* is not like learning the word *dog*. Conveniently, for dogs we can point to visually obvious examples, and a child can easily generalize from poodles to huskies and corgis. It is not even like learning what inner feeling to call thirst. *Conscience* is not only more abstract, but it has a social dimension: knowledge of community standards. Especially at first, a child will have only rudimentary knowledge of those. Moreover, learning social customs is often not explicit at all, but implicit, since we commonly mimic a behavior without really being aware of doing so.

As children grow up, they begin to appreciate that social contexts can be rather subtle, even when you have a passable grasp of community standards. Sometimes a kindly fib about someone's singing voice is better than the truth, and sometimes a well-intentioned offer to stack a neighbor's woodpile can be perceived as insulting their fitness. Some parents allow cursing and some forbid it. Social life is rife with subtlety: the things you can say or cannot say, and the best way to say the thing you normally should not say.

When does the word *conscience* typically enter into our conversation, be it inner voice or outer? By and large, it is when we are in a bind; for example, when the law requires one thing, yet conforming to the law would violate other strongly held values, such as truthfulness or fairness. As portrayed in Steven Spielberg's movie

Schindler's List, Oskar Schindler regularly broke the law by misleading his Nazi pals about the roster of Jews who worked for him in his factories in German-occupied Poland. To those who did figure out that the workers were Jews, Schindler offered bribes to buy their silence. Normally, of course, breaking the law—not to mention lying and bribing—is regarded as wrong, but sometimes that is what our conscience deems necessary.

We may consult our conscience when winning a game conflicts with fairness, as when a baseball pitcher considers beaning a star player with a fast pitch to put him out of the game. Or when one is tempted to sacrifice honesty for loyalty, as a staffer might lie to shield his boss from a charge of conspiracy. In 1975, White House counsel John Ehrlichman lied out of longtime loyalty to Richard Nixon—a decision he later came to regret when convicted of perjury.

There can be a discord between love of family and blowing the whistle on a sibling's crimes. Consider the inner conflict of David Kaczynski once he discovered that his brother, Ted, was certainly the deadly Unabomber who mailed bombs to scientists with the intent to kill them. Inform the FBI, or stay mum to shield your older brother? Fortunately, David chose the first option.

There can be tension between loyalty to one's friends and acquiescing to a law that you suspect is dreadfully wrong. Consider the playwright Lillian Hellman, who, when pressed by the House Committee on Un-American Activities in 1952 to name communist sympathizers in Hollywood, stood her ground, saying, "I cannot and will not cut my conscience to fit this years' fashion,"[7] while knowing her defiance would cost her dearly. It did. She was blacklisted and hence without work until the end of the decade. Many families know the turmoil of conscience when having to choose between prolonging a life of suffering and letting the life go peacefully.

In traditional discussions of what our conscience requires of us, conflict between choices is often the starting point. The medical staff at Memorial Medical Center in Uptown New Orleans during Hurricane Katrina (2005) realized they had to make some agonizing decisions. The hospital is 3 feet below sea level. During the hurricane, the hospital lost power, floodwaters were rising, the generators failed, and there was no outside help. Evacuation of the sickest was no longer possible. The staff had to apportion care, giving some patients priority over others. About forty-five patients died, some of whom might have survived longer but for the disaster.[8] Sometimes, we learn, there is no right thing to do—only the less terrible.

Such conflicts between values are part of social life—everyone's social life. Sometimes the option we finally settle on may be rooted in what we expect we can live with. Which may, or may not, coincide with its *moral rightness* according to our community's standards.

It is tempting to believe that our conscience can be tapped to deliver universal moral truths, and that as long as we heed our conscience, our choice will indeed be the morally right choice. The uncomfortable fact that has to be reckoned with, however, is this: conscientious people frequently differ on what their conscience bids them do, and hence differ in their choices. All too often there is a clash between what *your* conscience tells you and what *mine* tells me, even if we are siblings or neighbors or mates.

One of us may agree that genetic research related to racial features is important for medical development; the other is appalled by any such research. One of us may think that abortion following rape is acceptable; the other may hold that no justification for abortion is morally decent. Sometimes there is discord *within* my own conscience. Should I tell a prospective neighbor about a construction flaw I happen to know affects the soundness of the house they want to buy, or should I just keep silent? Is it any of my business? Why does my conscience not speak to me, loud and clear?

Not even a shared God ensures that our judgments of conscience will line up. As Abraham Lincoln observed, Americans in the South and the North read the same Bible, had the same God. Even so, the conscience of the southerners directed a course utterly opposed to that of the northerners.

Martin Luther confidently claimed that the Holy Spirit writes the moral truths on our conscience. Free of all misgivings, Luther claimed that the assertions of the Holy Spirit "are more sure and certain than life itself and all experience."[9] Realism intervenes: different devout hearts often deliver opposite moral assertions. Each may be stamped with certainty, but can both have God's blessing? Surely not. Worth recalling, too, is the multiplicity of world religions, among which significant moral disagreements are not uncommon. Buddhists may differ from Christians, who may in turn differ from Confucians. Christians differ among themselves. Heartfelt conviction is not, alas, a guarantee of moral decency. Self-delusional certainty, Socrates would remind us, is a symptom of our imperfection and the wobbly nature of our conscience.

The infallibility of conscience that Luther counted on, is unfortunately, a delusion, though that delusion often delivers the courage to plunge ahead with what our conscience urges us to do. In Martin Luther's case, or in the case of the abolitionist John Brown (1800–1859), we may find unwavering conviction admirable. But in the case of the Bolshevik leader Lenin, whose unflinching conscience endorsed the Red Terror in Russia (1918), or in the case of jihadists blowing up planes and kidnapping schoolgirls, we recognize the delusion as ruinous. Steadfast moral passion may be a good thing, but only when it is on the side of the angels. "And just when *is* that?" Socrates would surely ask. There is no answer that does not just go around in circles.

I may long for certainty, but I have to live with doing the best I can. I may concoct a myth to explain that *my* certainty, unlike yours, taps into universal moral truths. Reality will soon dissolve that myth. Voltaire (1694–1778), a French philosopher of the Enlightenment, concisely summed up the state of affairs: "Uncertainty is an uncomfortable position, but certainty is an absurd one."[10] Voltaire is right, of course, but nevertheless, we must make some choice or other, and not acting can be the worst of the options. Knowing that I am doing the best I can seems like cold comfort, yet my inner Socratic voice recommends that I make do with cold comfort.

Our most fundamental moral obligation, we are regularly admonished, is to act in accordance with our conscience. Is this truly what we should advise our children? Maybe yes, but then again, maybe no. Conscience is not always a trustworthy guide, because sometimes things turn out badly, confident judgments of conscience notwithstanding. Following a cultish ideology, someone might firebomb a theater or leak nerve gas in a train. When Thoreau tells us to act in accordance with conscience, it sounds reassuringly simple and straightforward, and yet, honesty bids us recognize that heeding his advice is anything but.

Obeying one's conscience is typically not a legal excuse for a crime. For example, Edward Snowden, a CIA employee who publically released top secret information, claimed to be goaded by a conscience troubled by the government's mass surveillance program. Regardless, Snowden has been held accountable for violating the Espionage Act of 1917. Were he to return to the US now, he would undoubtedly go to jail. Even so, what a jury would do with this case if Snowden returned not now, but in fifteen years or so, is difficult to predict. Which brings us to the next point. We know that our conscience can modify its assessments over time. As our lives go on, it is common to find that our attitudes have shifted regarding a social issue, such as whether the sale and use of marijuana should be legal.

Do we always find a way to rationalize our own wrongdoing, to make it seem all square with our conscience? Sometimes we do. What was going on in the conscience of tobacco company executives who knew full well that the scientific evidence showed a strong link between smoking and lung cancer, but who trashed the evidence and bought off politicians in order to continue marketing cigarettes? Where was the conscience of the priests when they were molesting altar boys? Whatever conscience is, it is not like the Earth's gravity, always pulling us in one direction.

So yes, conscience can be slippery, and misjudging a situation is as much a part of having a conscience as our conviction that we are getting it right. Despite all the messiness, however, many humans much of the time do try to be fair-minded and kind and honest, and particularly so within a group to which they belong, such as a family or a clan or a nation. We typically share and cooperate and help each other out of jams.

What accounts for that similarity in human behavior? What happens in the brain when we feel the urgency of a moral obligation to tell the truth or blow the whistle? What explains the troubled conscience when we knowingly choose to turn a blind eye to corruption? Can neuroscience help explain why we cooperate with each other, even with those we may not be especially fond of? Two skeins are apt to become twisted together here, and I prefer to keep them separate.

First, can science tell us, for any particular moral dilemma, what option our conscience should take—in other words, which is the morally right choice? No. Science cannot do that. Factual evidence can, however, assist with the decision. Science, along with all kinds of knowledge, can provide relevant facts—facts that decrease uncertainty regarding the consequences of an action, for example. Assembling the relevant facts can decrease the likelihood that we will eventually regret our choice. If you know that using a certain pesticide will result in the death of native bees that are needed for pollination, then that knowledge is relevant to determining whether to use the pesticide. If sex education for adolescents is known to reduce the number of unplanned teen pregnancies, that knowledge is a factor in figuring out whether sex education should be part of the school curriculum. Science can also weigh in on the social effects of many policies, such as making organ donation the default option on a driver's license, thereby increasing the availability of organs for transplant. Still, science itself does not tell us what is right or wrong.

The *second* question is very different from the first: can science help us to understand how it is that we are often motivated to care about what happens to others? Can it tell us how it is that we have a conscience at all, even if it cannot tell us what judgments our conscience ought to side with? Can it explain why you and I might differ in what our conscience tells us is right? Here I think the answer is yes.

We are, as Aristotle, Darwin, and many others realized, social by nature. We would have no moral stance on anything unless we were social. Fine, but what biological evidence supports the idea that humans are social by nature? Research in neuroscience has improved our understanding of what it is about wiring in mammalian brains, including human brains, that makes us social.

Very roughly, our infant brains are genetically set up to take pleasure in the company of certain others and to find separation from them painful. We are attached to mothers and fathers, to siblings and cousins and grandparents. As we mature, we become attached to friends and mates. These attachments are a profoundly important source of meaning in our lives, and they motivate a wide range of social behavior.

As infants grow and develop, they learn how their social world works. They come to appreciate how to play fair, work together, and forgive an insult. We learn by imitation, by trial and error, through stories and songs, and by reflection on our experience. We internalize norms concerning how to behave—sometimes consciously, sometimes nonconsciously. We acquire habits and skills for navigating the complexities of the social world we happen to be born into. Here, too, neuroscience has begun to understand what learning social skills and habits entails, in terms of the systems in the brain that change as we learn, and the genes that respond to make those brain changes permanent. One's own distinctive personality, such as being outgoing or not, also colors one's conscience, making it distinctive. As we will see in Chapter 5, my conscience may conflict with yours, owing to deep differences in basic aspects of our personalities.

Science itself does not adjudicate on moral values. When all available facts are in, we may still face the questions "What should we do?" and "How should we evaluate the facts to make the right decision?" Of course, individual scientists, as humans with moral values, may well have opinions on what should be done. Thus, many scientists, upon learning that there is a strong causal link between human papillomavirus and cervical cancer, campaigned in favor of vaccinating females against the virus.

It is not surprising that researchers who discover risk factors for a certain disease become eager to publicize that information, and eager to promote ways to reduce harm. Human papillomavirus is one famous case in point, but there are many others, including the discovery that smoking tobacco greatly increases the risk of lung cancer, that sharing needles may spread AIDS, that heavy alcohol consumption by pregnant mothers causes cognitive and motor impairments in babies. In all these examples, researchers worked not only as scientists bringing the data to public notice, but as concerned citizens wishing to improve people's lives. Scientists do this because, like other humans, they care.[11]

Needless to say, being a scientist does not automatically give you the moral high ground.[12] Humility, as Confucius (551–479 BCE) pointed out, is the foundation of all virtues. Often, but not always, laying out the facts will suffice to get agreement among the interested parties. Sometimes, however, the facts are complicated and do not just "speak for themselves." So a question can arise concerning whether to trust the information on the table.

Sometimes, because not enough is yet known, great uncertainty remains, even when the available facts are in. This can happen with experimental cancer treatments, where the data from clinical trials are not yet available. Terminal patients feel they should have the right to try an untested treatment, while researchers worry that an unfortunate outcome would set back progress that otherwise could be made. And sometimes fundamental values clash even when the facts are pretty well nailed down. There can be agreement on the facts concerning clear-cutting an old-growth forest, but disagreement on the moral value of conserving that forest versus logging a renewable resource. There can be agreement that a person is suffering terribly from a terminal disease, but disagreement about the value of doctor-assisted suicide.

In general, those who advertise themselves as having superior moral judgment or unique access to moral truth need to be looked at askance. Not infrequently there is great advantage—in money, sex, power, and self-esteem—in setting oneself up as a moral authority. The rest of us can easily be exploited when we acquiesce in these authoritative claims. Scam artists aplenty proclaim themselves as moral gurus, willing to tell the rest of us how our conscience should behave. They can seem authoritative because they are especially charismatic or especially spiritual or especially firm in their convictions. We will look into this issue again in Chapter 8. But the point belongs again to Confucius: humility is the solid foundation for all the virtues. Hence our suspicions are rightly aroused by moral foot stampers and moral blowhards.

Knowing that moral arrogance is usually a cover for manipulative intentions, Socrates queried the morally arrogant leaders of Athens, thereby causing them embarrassment. The leaders' answers to his polite but persistent questioning revealed that their self-assurance was backed by mere hot air. Their status as moral authorities turned out to be self-serving trumpery. Having been embarrassed, the city leaders were not pleased.

Accused by the prominent Athenian citizens of corrupting the minds of the youth by encouraging them to question authority, Socrates was condemned to death. The method?[13] Drinking poisonous hemlock. Devastated by the death sentence, Socrates's devoted students begged him to flee. He could readily have fled Athens with a customary but modest bribe here and there. Why he declined to flee has been much discussed ever since, as readers try to imagine themselves in his place, weighing the options.

Maybe the answer is as simple as Socrates said it was: he did what he thought was the right thing for him to do. Nothing fancier, nothing more subtle or existentially knotted up. Even so, as Plato describes the scene, with the numbness spreading up from Socrates's legs as the hemlock proceeds to kill him, we cannot but respect the resolute nature of Socrates's choice. Though the event was twenty-five hundred years ago, the story of Socrates's execution and its judicial pretext continues to be vividly relevant to our contemporary lives.

Between them, neuroscience and psychology have begun to explore how brains acquire values, including moral values, and how moral values guide decisions. If we think of conscience as involving the internalization of community standards, then one question concerns the processes that explain such internalization. Another question concerns how it happens that a community standard can change, or how an individual can come to regard a certain accepted social practice, such as binding women's feet to keep them small, as immoral.[14] The general outlines of the story, if not all the details, are now visible. It is not a simple story. So far, however, it is a coherent, biologically plausible story that can be told in a relatively simple way. It is a story that will change how we think about morality and ourselves as moral agents.

One caution needs to be flagged. Although neuroscience can help explain aggression in parents who defend their young against predators, aggression by one social group against members of another group is poorly understood at the level of the brain. What does seem evident from behavioral data is that ideology can be a powerful force in motivating such behavior, even when the odds are overwhelmingly against success.[15] The cynical hypothesis says that an ideological justification for out-group aggression is at bottom a way of buying off one's conscience for letting loose the exhilarating instincts of a predator. Young male adults may be especially prone to such a rationalization. In favor of the hypothesis is the typical dehumanizing language used to describe the enemy when one group stirs up resentment and hatred for another.[16] If the others are merely filthy animals, killing them is less odious morally.

Whether neurobiology eventually refutes or bears out the cynic remains to be seen. Important though such data are to us, their advent may be long in coming. The reason neuroscience has scant information on this matter is that gathering neurobiological data from humans engaged in deadly out-group hostilities is extremely tricky for all the obvious reasons. Devoted fighters in deadly combat are unlikely to be eager to take a break for a brain scan, for example. On the other hand, laboratory experiments that generate out-group hatred in groups of college students in order to detect and measure neurobiological properties thereby activated have a different kind of problem. They are ethically improper. What about rodent models of warfare? Although chimpanzees may infrequently engage in group hostilities, there are essentially no animal models of human warfare. So here is a topic where reliable brain data would be invaluable, but where getting such data is currently all but out of reach.

BEFORE WE PROCEED, I want to pick up a loose thread concerning definitions, as it will save us time and effort later. Psycholinguists have shown that our everyday concepts such as vegetables have a radial structure. This means that at the central core of the concept are examples that everyone agrees fall under the concept, but then encircling the agreed-upon core are examples that are somewhat similar to those in the core, but that not everyone agrees belong under the concept.[17] At the outer boundary, little agreement prevails about whether an example falls under the concept. The boundaries are thus fuzzy, not sharp. Instances of these very common concepts are *vegetable*, *friend*, *honest*, *house*, *river*, *weed*, *smart*, and so on. None of the categories have a precise definition, though they all have a definition in the dictionary that applies fairly well to the core examples.

The interesting thing is that we communicate very well most of the time, despite the data showing that lots of vagueness surrounds what counts as a vegetable or a friend or a house. Much of the time, the fuzziness of boundary cases does not matter. The data show that carrots are core examples of vegetables, parsley is out at the boundary, while tomatoes and squash are somewhere in between. The imprecision does not cause catastrophic breakdowns in communication. I have never needed to have a logic-chopping argument in the supermarket about whether parsley is really and truly a vegetable. Not once. Which is a good thing because there simply is no answer to whether the boundary cases "really" fall in the class or not. Moreover, psycholinguists point out that efforts at hewing out a precise definition for *vegetable* or *friend* do not, in fact, lead to greater clarity in conversation, but instead lead to nonproductive wrangling, and anyhow people go on speaking as they did before. The vagueness of the boundaries is generally not a problem, but an advantage that probably makes linguistic change possible as speakers extend the use of a word in a novel but useful way.

In a legal context, on the other hand, key concepts frequently are given reasonably precise definitions, such as the minimum age for getting a driver's license, or what counts as *driving under the influence*, which in most states in the US means having a blood alcohol content of .08 or greater. Some adolescents are competent and responsible enough to drive a vehicle at age fourteen, and others are not ready even at twenty-two, but a uniform policy is needed, so in most states, sixteen it is.

Even in the law, however, legally honed concepts may still have unavoidable vagueness. For example, the law specifies *negligence* as that which deviates from the conduct of a person of ordinary prudence. But *ordinary prudence* itself is not precisely defined. Despite the imprecision, the law generally works, since most speakers of the language understand well enough what is meant. As a concept with clear central cases and fuzzy boundaries, *morally right* is more like *ordinary prudence* than like the legal definition of *driving under the influence.*

In a scientific context, too, efforts are made to provide precise definitions of certain category names, such as *planet* or *protein.* Even here, however, scientists typically manage with a rough-and-ready characterization until the data are sufficient to introduce more precision. A good case in point is *gene,* which was only crudely delimited as "the carrier of hereditary information about a trait" before 1953, when James Watson and Francis Crick discovered the structure of DNA. The meaning of *gene* has become increasingly precise over the last seventy-five years as molecular biology has made more discoveries concerning how DNA codes information, and how that information is used to make proteins. Before 1953, no one could define a gene in terms of sequences of DNA because not a single person knew anything much about DNA. In the same vein, before the middle of the eighteenth century, no one could give a precise definition of *fire* (now known to be "rapid oxidation") because no one knew there was such a thing as oxygen or a process such as oxidation. Notice, however, that people managed to talk about and investigate fire quite well nonetheless. Definitions evolve as scientific discovery enables additional precision. Scientific definitions typically emerge at the late point of discovery; they are not necessary or even possible at the beginning.

This aside is useful for what follows because crucial concepts under discussion, such as *conscience, moral, decision,* and so forth, cannot be precisely defined at this stage of our inquiries. As with all our everyday concepts, these have a radial structure with a central core and fuzzy boundaries. Additional precision has been possible for *decision,* owing to striking discoveries in neuroscience over the last decade concerning how neurons and neural networks integrate information from various sources to make a decision. These discoveries turn out to be relevant to moral decision-making, and they yield a deeper understanding of the nature of morality in human societies.

CHAPTER 1

The Snuggle for Survival[1]

A mother's love for her child is like nothing else in the world. It knows no law, no pity, it dares all things and crushes down remorselessly all that stands in its path.

AGATHA CHRISTIE

SOMEONE TO LEAN ON

Lizards and garter snakes live a solitary life. Not so wolves or gorillas. Or humans. We are intensely social. We take pleasure, great and small, in the company of friends and family.

Loneliness is stressful, whereas re-joining loved ones is joyful. We form strong and enduring bonds both within a family and with friends outside of it. We tolerate—and are attached to—annoying offspring, infirm parents, and pesky neighbors. We find exile deeply distressing; a long stretch of solitary confinement is a particularly devastating form of punishment.

Variability in how much close company we want is also part of our social experience. Some of us are introverts, some extroverts, and many ramble around somewhere in between. Our desire for company varies with age and life experiences, but very few seek the completely solitary life. Fur trappers of yesteryear did manage to plod on alone for long months through dark winters. Even then, however, they often had a dog to keep them company. Spring brought the raucous rapture of reuniting with other humans.

I recently visited a deconsecrated monastery in Tuscany (opened in 1343) where the previous inhabitants had been Carthusian monks. A primary principle of the Carthusian order is that the monks are, one and all, hermits. Even these hermits, however, found value in living with other hermits in the same monastery, attending mass together, defending each other against bandits (a common problem in the fourteenth century), and baking bread in a communal kitchen. Stacked up against our strong inclinations to "tend and befriend," the variable preferences for a bit of solitude now and again seem rather modest.[2]

Living in a community normally boosts one's chances of surviving and thriving. We can share food and huddle against the cold; we can organize to attack prey or to defend each other against invaders. We can divide the labor of making a living, allowing for the emergence of distinct kinds of expertise, such as herding goats or blacksmithing iron into useful shapes. Although unpredictability in weather, food resources, and plagues can swiftly devastate a community, these troubles can be modulated by ingenuity. Boats can be built, tools invented, and animals domesticated. Vaccines have truncated the swath of death caused by infectious diseases such as smallpox and polio. Vaccines are the product of a social community where truth is valued, knowledge accumulates, and cooperation prevails.

Despite the many benefits of sociality, our wider social world sometimes snuffs out the opportunities to thrive. Wars scorch the Earth and leave its inhabitants maimed and massacred. Not unusually, history shows, one band of humans enslaves another. In an otherwise flourishing community, corruption may undermine trust, which undermines cooperation, smashing the institutional scaffolding that holds people together. Then they may be at loggerheads, and civil war ensues.

If our sociality motivates caring for others, it is also true that we are given to hate. We humans regularly derive pleasure from hating those we consider outsiders. We tend to find hating energizing. When things go awry in our lives, we can be perked up by hating and blaming outsiders or misfits. Hating those we regard as strangers may strengthen the bonds within our inner circle, which itself can make us feel elated. Our self-esteem soars as we tell each other how superior we are to those miserable wretches in the other group. Adolescents in gangs can be motivated to destroy for a lark: vandalize a church, torch an old woman's shed, set a boat adrift—mayhem is not beyond our ken.

Hell-raising antics aside, why are we socially driven at all? Unlike lizards, we are attached to our family and friends. We want to hang out with them, and we yearn for them when we are separated. We care deeply about whether they like us. We work on common projects and help each other solve problems. In some situations, we go out of our way to assist strangers. What is it about our mammalian brain that makes us like that? We never see a turtle help out a struggling salamander, but we are not surprised to see a dog befriend an abandoned kitten.[3]

The fast answer is that the brains of our mammalian ancestors were adapted for sociality. The adaptation involves a trick we often see in biological evolution—namely, the repurposing of an existing function to yield something rather new that happens to be advantageous in the struggle to survive. A few genes get altered or duplicated, with the upshot that an old function gets a new look and a new application. In the evolution of the mammalian brain, feelings of pleasure and pain supporting self-survival were supplemented and repurposed to motivate affiliative behavior. Self-love extended into a related but new sphere: other-love. If mammalian sociality was favored by evolution, what exactly was the benefit and which *others* mattered? The answer turns out to be less than obvious, and requires a somewhat roundabout account.

BORN HELPLESS

The first and most basic evolutionary point is that the primary targets and beneficiaries of sociality are the offspring. Why? Because mammalian babies are immature at birth and will certainly die without care. Baby turtles, after hatching from their eggs, immediately dig their way up out of the sand, scuttle down to the water, and begin to look for food. No parents are anywhere close by, nor are any needed. Baby rats, by contrast, are deaf and blind, with no fur to keep them warm. Their skin is so thin that you can see the gut through it. As for finding food, the rat pups' bare minimum of innate reflexes help them snuffle around, find a warm protrusion, and then suck on it. With luck, it is something that yields milk.

An evolutionary complication for all mammals and birds with helpless offspring is how to arrange for care. No animal will just volunteer for this demanding and time-intensive job, so adaptations are needed to motivate offspring care. The only viable candidate to care for newborns? Mothers.[4] Mammalian mothers are close by when the live babies are born, unlike iguana mothers, which lay their eggs and are long gone by hatching time. Mammalian fathers may not be close by either, unless yet other genetic changes ensure that mothers and fathers are bonded to each other. And that is another story. In addition to body modifications, such as a womb, a placenta, and highly nutritious milk, changes occurred in mammalian brain circuitry to guarantee that mothers care for their helpless newborns.

All animals must have the basic circuitry for self-care, or they will fail to survive long enough to reproduce. In the evolution of the mammalian brain, the range of *myself* was extended to include *my babies*. Just as a mature rat cares for her own food, warmth, and safety, so she cares for the food, warmth, and safety of her babies. Genes build brains, and new mammalian genes built brains that felt discomfort and anxiety when the babies were separated from their mother, such as when they were snatched out of the nest. On the other hand, the mammalian brains felt calm and good when the babies were close by, warm, and safe. These new mammalian brains felt pleasure when they were together with their babies, and the babies felt pleasure when they were snuggling up with their mothers.

Physical pain is a "protect myself" signal, and pain signals lead to corrective behavior organized by self-preservation circuitry. In mammals, the pain system is expanded and modified; protect myself *and* protect my babies. In addition to a neural pathway that identifies the kind of pain and locates the site of a painful stimulus, there are pathways responsible for emotional pain, prominently associated with the cortex, but also with older structures below the cortex. One special cortical area, the insula, monitors the physiological state of the entire body. When you are gently and lovingly stroked, this area sends out "safe" signals (doing very well now). Such stroking is known as affective touch. The infant responds likewise to gentle and loving touches: ahhhhh, all is well, I am safe, I am fed. Safety signals downregulate vigilance signals controlled by stress hormones.[5] When anxiety and fear are downregulated, contentment and peacefulness can take their place. These social feelings are the foundation for attachment, and with contented snuggling, the bond between infants and mothers intensifies over time.

While mammalian mothers typically go to great lengths to feed and care for their babies, a mother's tending of her young may interfere with feeding herself. Moreover, danger lurks when predators detect dinner potential in one's young. These costs to the mother explain why the baby-care wiring evolved to be highly robust against minor discomforts and even major dangers. Mother love is a mighty force, not a mild inclination. Rarely do mammalian mothers cavalierly abandon their young, and when they do so, it is usually because something has gone very wrong in the maternal brain's wiring.

How does the new mammalian wiring work to drive baby care? The answer is not entirely known, but some crucial facts have been discovered. There are four key microplayers in the neurobiological drama supporting mammalian infant care. Their action can be extended as care is extended to mates, kin, and friends. The first two are the neurohormones oxytocin and vasopressin. The third and fourth are the opioids and cannabinoids that your brain makes and that cause you to feel good. This quartet stands out against the orchestral background of sex hormones—estrogen and progesterone—and yet other neurochemicals, such as dopamine, that enable the mammalian brain to learn from experience. More detail is needed to amplify this preliminary account, and that will be the aim of the next chapter.

For now, however, we are focusing on why evolution favored rewiring for baby care, so the pressing question now is why mammalian and avian babies are born so helpless. If newly hatched turtles are precocious—mature enough to be independent at hatching—why not neonate mammals and birds? *Altricial* infancy (being born helpless) seems almost like a retrograde step in biological evolution. Was it an evolutionary error, or were there benefits to neonatal immaturity? Yes indeed, benefits abound.

BORN TO BE WARM

The explanation for altricial infancy is rooted in the unique urgency for mammals and birds to be smart. The story starts in an unexpected place: with the emergence of *endothermy*, which is the capacity to generate one's own constant body temperature, regardless of temperature fluctuations in the environment external to the body. Being warm-blooded, in other words. When endotherms first appeared 250 million years ago, they were reptilian and small—not yet mammals in the true sense. Endotherms enjoyed a masterful advantage over their cold-blooded competitors (ectotherms): they could forage even at night, without the warmth of the sun. Nobody else was there. As biologists put it, the endotherms invaded the *vacant nocturnal niche.* Perhaps they even fed on sluggish exothermic insects awaiting the sun's warmth to energize them. Easy pickings.

Importantly, the endotherms could forage without fear of competition from cold-blooded dinosaurs. Endotherms could also thrive in colder climates, thus opening new feeding and breeding ranges that were closed to their cold-blooded cousins. Endothermy was a very, very big deal. Over long evolutionary time, it triggered a set of interlocking changes that produced very smart, social animals—mammals and birds—that are strongly motivated to care not just for themselves, but for others. In highly social mammals such as humans, marmoset monkeys, and wolves, there is care not just for babies but for mates, kin, and friends. And sometimes for individuals of other species, such as dogs and goats.

The earliest warm-blooded species eventually became extinct, as furry mammals appeared and became ever more successful and ever more intelligent. How could the capacity to generate one's own heat lead to cleverness and to sociality?

Life has its trade-offs. Wonderful though the warm-blooded advantage was, it came with a major cost: gram for gram, an endotherm must eat ten times as much as a cold-blooded creature in order to survive.[6] Lizards can go many days with no food, but rats would starve to death in similar conditions. Energy requirements on this scale amount to a formidable ecological constraint. If you cannot get the calories you need, you become someone else's calories. What changed in the warm-blooded brain to cope with the exceptionally high demand for calories? Being smarter.

BORN TO BE SMART

In a cutthroat world, being smarter than the competition is an advantage, other things being equal. What does being smart mean, in this context? Mainly, it means you have an enhanced capacity for understanding your environment and applying that knowledge in foraging, mating, and surviving. It means you can make finer distinctions among stimuli (crackable nuts versus uncrackable ones; healthy versus unhealthy potential mates). It means you can figure out causal relations among distinct but similar kinds of things (edible insects versus insects that bite back). Upgrading the capacity to learn about the world is an efficient path to smartness.

An alternate—nonlearning—route to enhancing smartness would depend entirely on genetic mutations, emerging over massively long timescales. With luck, the mutated genes would construct brain circuitry to embody enough world information so that the organism could usually survive and reproduce. This is essentially how simpler organisms like frogs manage as well as they do. If endotherms had had to wait for such genetic mutations to crank up their intelligence game, they would have been wiped out.

The further limitation to the wait-for-mutations strategy, apart from its dependence on extremely long timescales, is that built-in knowledge lacks flexibility, should the world happen to change—something the world is prone to do. After all, if the genes build into the wiring the knowledge that rabbits are what you must eat, that is fine when rabbits are plentiful but a drawback when rabbits are scarce. You may end up unsuccessfully scrounging for rabbits and ignoring turkeys and trout, plentiful and nourishing though they are.

In mammals and birds, learning was scaled up in an exceedingly big and genuinely new way. In organisms such as cockroaches and frogs, learning mechanisms are limited to small adjustments to the circuitry that is largely controlled by instincts. Mammals, by contrast, are *Big Learners.* After mammals are born, their brains grow as much as fivefold, as they build ever-more-intricate patterns of connections among neurons, with the result that genetic control of behavior is buffered by learned control. Big Learning permits the formation of long-term plans, along with intelligent evaluation of distinct options based on knowledge of cause and effect in the environment. Although the genetic foundation for biasing behavioral decisions never disappears in any species, it can become less and less dominant as the capacity for learning increases. The knowledge structures built on the instinctual foundation can be modest, as they are in mice, or cathedral-like, as they are in humans.

Flexibility means you can change when the world changes. Tight genetic guidance of all aspects of behavior can become an impediment when a world shifts or a new environment is explored. For example, cockroaches do very well in Fiji, but not in Alaska, whereas humans—and rats—do well both in Fiji and in Alaska, vast environmental differences notwithstanding. The upshot is that a powerful learning platform is favored by evolution. There is a snag, however, for boosting the learning capacity to enable a very high degree of flexibility. That snag is the immaturity of the nervous system at birth.

Why are Big Learning and immaturity at birth interlocked? A basic neurobiological fact about learning provides the answer. For learning to occur, a structural change must occur in the brain to encode what is learned. More specifically, individual neurons in the appropriate network must change their structure a bit, yielding a difference in the architecture of the network. This structural change is, in effect, an embodiment of the learned knowledge. A neuron might change by adding a contact to other neurons—that is, by making a new synapse or two (Figure 1.1). Or a neuron can also expand its input branches or its output branches. On the other hand, sometimes branches that are not doing much work are pruned back, making more space for new growth on highly active neurons.[7]

To maximize the impact of experience on the neuronal networks, the wiring itself must be as minimal as possible at birth, yet sufficient for maintaining life outside the womb. Why? Because neurons need room to sprout and spread if they are going to encode what has been learned. If neonatal neurons are already in a mature state, they are already committed by genetic determination to their function in the network. Consequently, not much structure can be added without compromising the

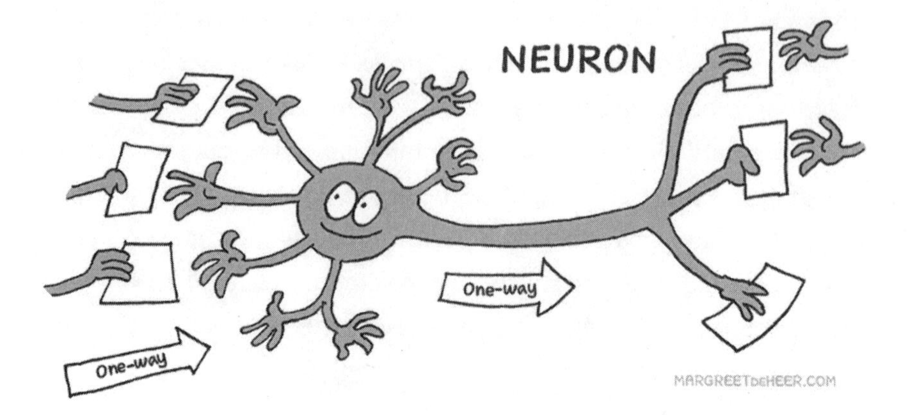

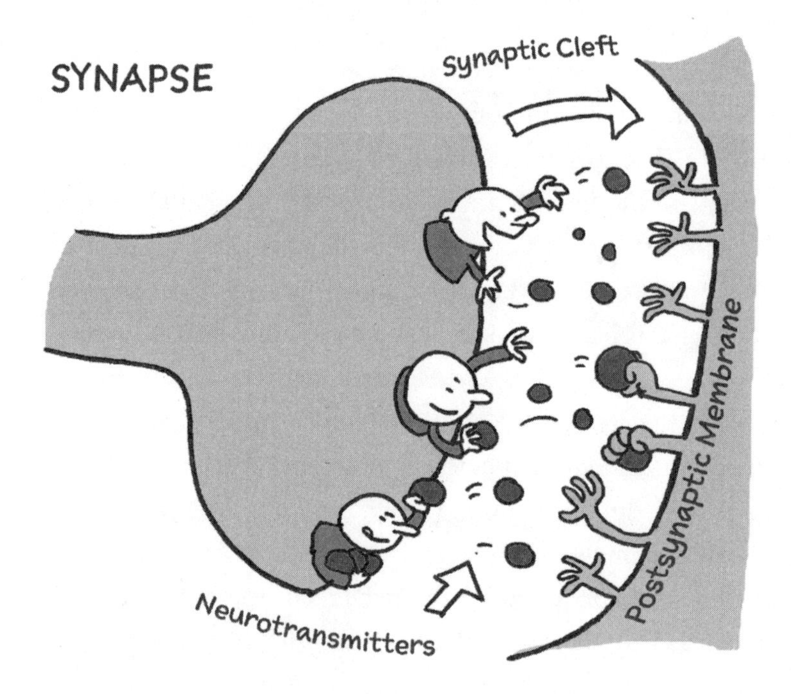

Figure 1.1 Cartoon showing the main structures of a neuron. Top: Input signals enter mostly on the dendrites and cell body, where signals are integrated. The output travels down the axon to the synapse. Bottom*:* The synapse is a point of communication between neurons. The axons release neurotransmitter, which drifts across the space between the neurons (synaptic cleft), and binds to specialized receptors on the receiving side. This communication by neurotransmitter will change how likely it is that the receiving neuron is activated. A cortical neuron may have between one thousand and about ten thousand synapses. COURTESY MARGREET DE HEER.

instinctual responses that the genes organized the neurons to perform. If new synapses and new growth are curtailed, then causal knowledge about how to thrive in the world is also curtailed. At birth, therefore, any brain destined for flexible smartness must be immature. Mammalian babies are Big Learners and thus altricial.

True enough, there are *precocial* mammals such as goats and bison, where the newborns are typically able to stand and walk very soon after birth. Those that cannot are doomed. Such precocity, however, seems to have been a much later adaptation in mammalian evolution, occurring mostly in hooved, plant-eating species of herding mammals. Moreover, the brains of these precocial babies do not display the great growth typical of altricial babies.

Even in precocial mammals, however, significant calorie dependency is still the norm in neonates. Kids suckle from the nanny goat for two or three months; bison calves suckle for about six to eight months. Their precocity seems to be largely limited to their capacity to walk and suckle from a standing position. Bear in mind, too, that magnificent though they are, bison are not nearly as clever as wolves or raccoons. They spend much of their days munching plentiful grass, and for such defense against predators as is needed, they rely mostly on the statistically favorable advantage of being one among others in a huge herd. A bear or mountain lion might venture onto the open prairie to go after a straggler, but those safely situated centrally in the herd typically do not show more than modest concern about the unfortunate straggler. A nearby bison mother might charge and kick, but a wily bear can usually outwit her.

BORN WITH A CORTEX

To understand more deeply how being smart and being social came together in a unique way in mammals, we need to ask, How did the brain wiring change during the evolution of mammalian brains, and how did that wiring enable the power and flexibility that we associate with intelligence in monkeys and wolves and us? That is, how did it eventually enable such sophisticated capacities as complex problem solving, self-control, imagination, and conscience?

Cortex. That is the crux of the answer. The cortex is a brain structure that is unique to mammals.[8] All mammals have it; no mammalian ancestors do.[9] Were you to look down through my skull, the neural structures you would see are the hills and valleys of my cortex, overlying but highly interconnected with, evolutionarily more ancient structures underneath it.[10]

The architecture of cortex is utterly distinctive: six neatly stacked layers of neural circuitry, with specific neuron types precisely located in their designated layers, making prototypical connections to other neurons (Figure 1.2). The cortex has essentially the same architecture in all mammals, whether bats or baboons or us. Within an individual brain, cortex has essentially the same striking organizational profile in all areas, whether the area is specialized for processing visual signals or auditory signals, or for organizing finger muscles to accomplish a task such as threading a needle. Cortex was the structural innovation that ushered in Big Learning, which in turn enabled mammals, with their high calorie needs, to succeed in the world.

Strictly speaking, the general term *cortex* applies to any neural structure that is organized into layers—in other words, that has a laminar structure. The contrast to cortical structure is *nuclear structure*, meaning, roughly, "a clump"; the places where neurons receive and send signals are aggregated into clumps, rather than nicely laid out in layers. An example is the *nucleus accumbens*, a subcortical structure that plays an important role in attachment. Its front end mediates pleasurable responses, while the back end mediates dread and disgust responses.[11]

The hippocampus, an ancient structure predating mammals and crucial for spatial memory in all of us, has a highly distinctive three-

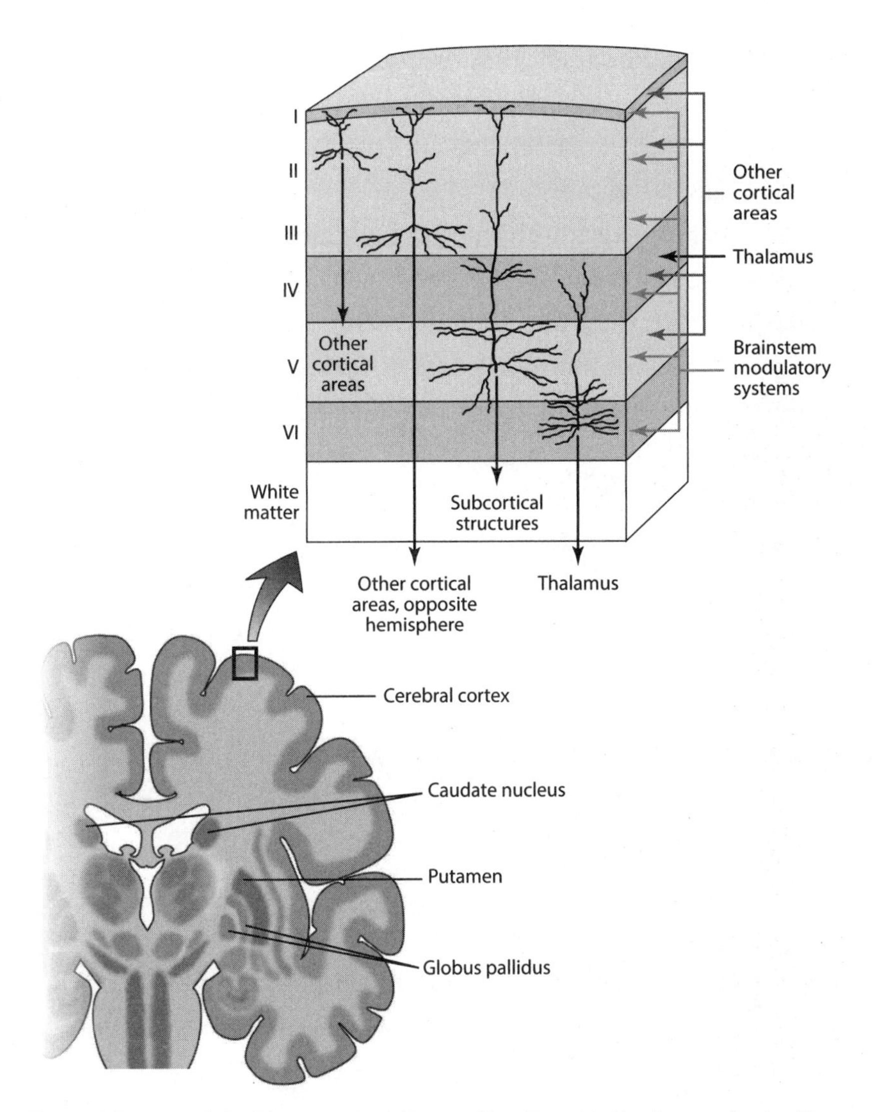

Figure 1.2 Lower left: Diagram depicting a slice through the human brain. The dark-gray edge on the surface is the cortex. The holes situated more centrally are the ventricles, and are filled with fluid. The dark regions below the cortex are the various subcortical structures, such as the basal ganglia—globus pallidus, putamen, and caudate nucleus. The white areas between the cortex and the subcortical structures are densely packed axons of neurons, carrying signals from one region to another. The axons of neurons are white because they are wrapped by fatty insulation (myelin) that is lighter than neuronal tissue in the gray matter that lacks myelin. The percentage of oxygen used by gray matter is about 94%; for white matter, it is about 6%. Upper right: Schematic of the six cortical laminae. The diagram indicates the specificity of input sources and output targets for neurons located in distinct laminae. Neurons are densely packed, about 100,000 in a cubic millimeter of human cortical tissue. MODIFIED WITH PERMISSION FROM THE *ANNUAL REVIEW OF NEUROSCIENCE*, VOLUME 26 © 2003 BY ANNUAL REVIEWS, HTTP://WWW.ANNUALREVIEWS.ORG.

layer organization. It is called *archicortex* to reflect its ancient origins. The six-layer cortex that is unique to mammals is sometimes referred to as *neocortex*, to distinguish it from premammalian two- or three-layer structures.

Laminar organization has engineering advantages. For one thing, it maximizes neuronal connectivity while minimizing lengths of axons and dendrites, thus keeping wiring costs down. For another thing, laminar organization provides a kind of scaffolding so that specific operations occur where they contribute appropriately, maybe even optimally, to the brain's ongoing computations.

Nevertheless, the caution to consider is that the neural architecture in birds, which are both social and smart, is largely clumpy. Bird brains do not have the six-laminar cortex that is typical of every mammalian species. As it turns out, they can be very smart anyhow, as we know from the clever behavior of bird species such as ravens and parrots.[12] This anatomical contrast between the brains of birds and brains of mammals suggests that about 150 million years ago, when birds separated from dinosaurs, evolution stumbled on a neurobiological innovation to upgrade intelligence that is somewhat different from that found in mammals.[13]

One striking feature of cortical circuitry in mammals is that it is scalable. Mice have a small cortex, monkeys have a much larger cortex, and human cortex is much larger yet again (see Figure 1.3). Species differ in the proportion of cortex that supports processing of signals in a particular modality. Rats have a small cortex for auditory processing, whereas bats, which use echolocation to navigate in the dark, have a huge auditory cortex. Monkeys and humans have a very large proportion of cortex devoted to visual processing, whereas naked mole rats, which live exclusively underground, have almost none.

These differences in cortical specialization in different species notwithstanding, the organization of neurons in cortex is basically the same. Human cortex is distinguished mainly by having a greater number of neurons, and hence is bigger than the cortex of other primates. The orderliness of the canonical circuitry of the cortex probably is what makes it scalable, since the genes for building cortical tissue in the fetus can just be turned on for a more extended period, and the new additions fit right in with the existing circuitry. The scalability of cortex also suggests that the genetic adjustments involved in producing additional neurons to make additional cortex in a species readily occur.

Importantly, the genetic portfolio and the principles governing cortical development in the embryo and the infant appear to be widely shared among *all* mammals.[14] This means that the roughly 200-million-year-old cortical innovation worked well in its early days, and it still works well now. There are some differences between the cortical genetics of mice and primates.[15] One interesting modification is that individual neurons are *much* smaller in primates than

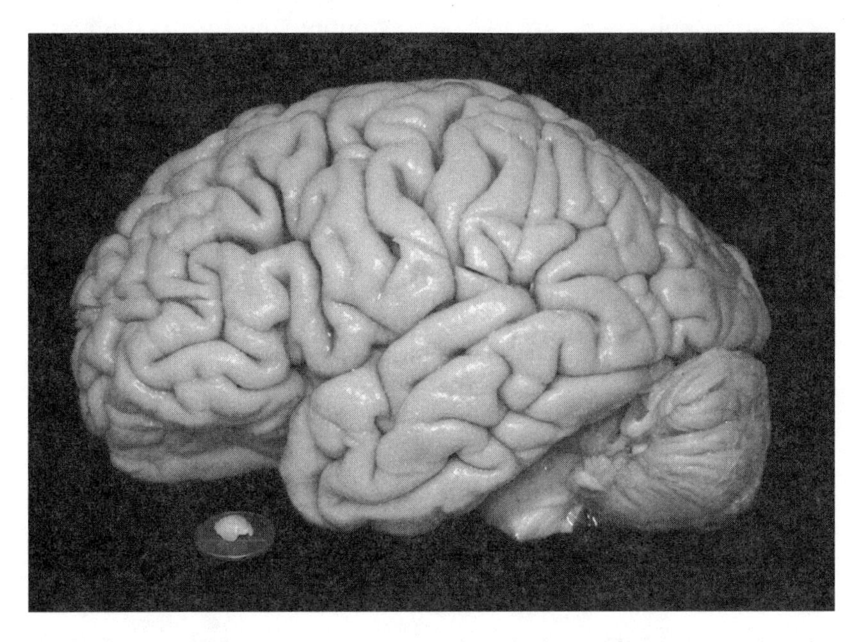

Figure 1.3 An adult human brain compared to the brain of a shrew (lower left), which has been placed on a penny for scale. COURTESY K. C. CATANIA. SEE ALSO K. C. CATANIA, "EVOLUTION OF THE SOMATOSENSORY SYSTEM—CLUES FROM SPECIALIZED SPECIES," *EVOLUTION OF NERVOUS SYSTEMS* 3 (2007): 189–206.

in rodents, with the upshot that you can pack far more neurons into a cubic millimeter of the canonical circuitry of primates.[16] Miniaturization of neurons is an adaptation of primates. Mice have only about 14 million cortical neurons, which can comfortably fit into a mouse's small skull. A monkey, however, has about 2 *billion* cortical neurons, while a human has 16 *billion* cortical neurons, so unless their neurons are much tinier and much more densely packed than those in the mouse brain, a disastrously gigantic head would be necessary. Miniaturizing processing components is something that computer engineers understand well.

Although the evolutionary origin of cortex is not well understood, it is probable that in very early mammals, enhanced senses of smell and touch increased foraging success, because these enhancements facilitated navigating in the dark of night. For species that eventually became diurnal, eyes that yielded good information in full sun, dusk, and dark, were also highly advantageous.

At some point in neurobiological evolution, these genetic changes to sensory modalities began to link up with an emerging neural structure that could efficiently integrate diverse kinds of sensory signals to extract high-grade information, the better to guide foraging and safety decisions. For example, in foraging it is advantageous for a brain not merely to register lower-grade signals such as "motion there," but to integrate signals from smell, touch, and vision to get a more specific and informative signal, such as "edible, fresh cricket there" or perhaps "nasty, inedible cricket there." Foraging-specific information also saves time and energy.

From an engineering perspective, the highly regular neuronal organization of cortex is well suited to integrate diverse signals and to pull out abstract representations of events and things in the world that are relevant to survival and reproduction. As far as we can determine now, the more cortical circuitry that is added, the more effective is the capacity to discern complex causal patterns in the world.[17]

A major part of the magic of the cortex is that it learns, integrates, revises, recalls, and keeps on learning. Infant human brains make about 10 million synapses (neural connections) each second. By puberty, the human brain has increased its weight fivefold over its birth weight. Learning about the physical and social world reached a new degree of sophistication with the emergence of cortex and its subsequent expansion in many species, including hominins such as our ancestors *Homo erectus* and *Homo neanderthalensis*.

Some genes play a crucial role in laying down the basic wiring during fetal development, while certain other genes have a role after birth in regulating protein production during the vigorous growth of neuron branches to support learning.[18] To build circuitry that can model regularities and values experienced in life, the brain needs to make proteins that become part of the very structure of sprouting neurons.[19] This is how enduring memories are made. Consequently, genes that assemble new proteins to make new neural components must be expressed during learning.[20] Plasticity on a big scale is part of our nature.

The nature of the cortex is to modify its connectivity so as to map the effects of nurture, to put the matter in the context of the nature-versus-nurture cliché. This is its genius. This capacity of the cortex to engage in Big Learning is what enables the flexibility seen in mammalian behavior. The neuron-intensive, six-laminar architecture provides the power to model species-relevant properties of the world. This flexibility and power are two elements that make for what we think of as intelligence.

Tuning up the brain to map features of the environment can be highly advantageous, as long as the cortex is interlaced with a system for assigning value to things (dangerous or safe, tasty or nasty). Absent the highly organized connections of cortex to ancient structures such as the basal ganglia, which are crucial for motivation, valuation, goals, and emotions, the cortex would be pretty much useless—a hood ornament. These ancient structures are the headwaters of motivation and drive, lust and hunger and thirst, and movement sequences. They regulate sleep, waking, and attention shifts. Even one's conscience is not solely, and perhaps not even largely, a cortical function. Social intelligence does indeed depend on cortical functions, but it crucially depends also on evolutionarily more ancient structures, such as the basal ganglia. These subcortical structures play an essential role in valuation.

The frontal regions of the mammalian cortex connect to subcortical structures that include the basal ganglia, and these connections support the process of learning what to approach and what to avoid. They also enable value-based judgment, such as whether to suppress an impulse. For example, a basic lesson for survival in the bush is never to run from a bear. Bears can run much faster than any human, even Usain Bolt, and they are eager to chase if you run.[21] Standing still despite a powerful impulse to run requires an unholy effort in self-control, but humans can do this. During the encounter, your frontal cortex will be highly engaged with your basal ganglia, as it draws on its acquired skills to suppress instinctive behavior that would be deadly.

Exactly how the handshake between old basal ganglia and new cortex was achieved is not well understood. That it *was* achieved is clear from comparing the anatomy of mammalian and reptilian brains. Although it has been suggested that under our fancy cortex sits an old reptilian brain, this should be taken as only a playful comment on the evolutionarily ancient origins of subcortical structures such as the basal ganglia. In reality, under my cortex sit basal ganglia that have a homolog in lizards, but that are distinctly mammalian and highly integrated with my cortex. My basal ganglia could not function in a lizard.

BORN FAMISHED

In addition to requiring a caregiver for the offspring, the Big Learner faces another big challenge. Brains are energy hogs, because neural activity in integrating and sending signals requires energy. The human brain is about 2% of our body mass, but it uses about 25% of our calorie intake.[22] The human brain has about 86 billion neurons, which means that about 6 calories per billion neurons (about 516 calories) per day are needed just to feed our brains. Consequently, the calorie intake of endotherms scales way up not just to keep body temperature within survival range, but also to keep their smart brains smart.[23] The energy needs of neurons limit brain size, given that other organs (heart, lungs, intestines), as well as muscles, need energy to operate. From time to time, evolution favors a dumbing down; sacrifice some smartness in exchange for larger skeletal muscles, like bison, or for a digestive system that can ferment grass, again like bison.

Calorie pressures do not end with endothermy or with providing energy to keep the existing neurons functioning. Immaturity at birth calls for especially high calorie intake to build the new brain circuitry to embody what is learned through experience. Another energy-related problem is that immature digestive systems are not set up for adult food. Although mother's milk is perfect for mammalian babies, the mother must find the extra energy resources so that her body can make rich milk to ensure that her potentially smart offspring get the calories they need to build their brains. If the mother does not get sufficient calories, the babies suffer malnutrition, which results in brains that are cognitively disadvantaged.

As an aside on calories, it is interesting that in some species the mother, having just given birth, will eat not only her placenta (afterbirth), but also any newborns that are defective or deformed. This behavior has been seen in numerous species, including black bears, rodents, and primates. Although this practice of cannibalism may strike us as shocking, the fact is that such consumption provides rich protein food that helps the mother's calorie quotient, reduces her feeding obligations, and provides richer milk for those babies likely to thrive.[24] Why human mothers do not usually eat the placenta is unknown.[25]

Other way-of-life changes in mammals and birds are shaped by bioenergetic constraints. Because a mammal must eat so much more than a similar-sized reptile, a given patch of land supports fewer mammals. Dozens of lizards can feed quite well on a small patch, but a patch of such size will support fewer squirrels and even fewer bobcats. One implication is that evolution cannot favor provisioning larger brains unless there is an accommodation—namely, smaller litters. Accordingly, one successful adaptation consists in producing fewer rather than many offspring, and in investing heavily in their welfare until they reach independence. Eight in a litter of rat pups may seem like a lot to a human, but it is small relative to a garter snake litter, which can number around fifty to ninety.

Calorie constraints crop up for other reasons. In the evolution of the mammalian brain, mothers are typically wired to defend their young, sometimes ferociously. In some species, such as prairie voles and wolves, the fathers will also vigorously defend the young. Defense against predators is taxing and requires a lot of energy. The thing is that, by the time the offspring are big enough to make a sizable meal for a predator, these young'uns constitute a significant energy investment by the mother. For species that bear just one or two offspring after a long pregnancy, every offspring represents a massive investment. To a first approximation, evolution favored mothers who were wired for spirited defense, as well as for recognizing when a cause is lost and the best action is to abandon the young and try again another day—sometimes a subtle call.[26]

Greater brain size relative to body size boosts smartness, but in larger-brained species, the mother tends to have a lower production rate; that is, the spacing between births is longer for animals such as chimpanzees and humans than it is for rats and mice. Roughly, the greater the number of neurons in a brain, the longer the period from birth to maturity. The longer maturation period is owed primarily to the energetic demands of Big Learning; having many more immature neurons means needing vastly more calories. Rats have tiny brains compared to chimps and humans. Baby rats are ready to leave the nest at about twenty-two to twenty-four days after birth, and they reach sexual maturity at about sixty-five to seventy days. By contrast, a chimpanzee baby nurses from its mother for about five years and stays with her for about ten years, and females have their first baby at about thirteen or fourteen years.

The constraints on production rate can, however, be relaxed a little. How? If mothers get help. Suppose others, such as a mate, bring the mother food, carry the young, or guard the nest while the mother is foraging. Biologists call these activities *energy subsidies*. Then, keeping brain size constant, she can give birth more frequently.[27] Human mothers lucky enough to enjoy generous energy subsidies can give birth more frequently than their smaller-brained chimpanzee relatives do—every two or three years.

Energy subsidy of mothers is typical in species where both parents rear the young, such as humans, marmoset monkeys, and prairie voles.[28] In these species, the fathers, and sometimes the older siblings, help the mother, perhaps just by carrying the infant while she forages, as in titi monkeys. Cooperative mothering, where many females assist other females, has been seen in several species of capuchin monkeys, in baboons and mouse lemurs, and in humans. Allo-suckling of one another's pups in a shared nest is common in house mice, and does not necessarily involve kin.[29] In chimpanzee troops, female dyads are common, in which each female helps the other with infant care,[30] but close analysis suggests that the scale of this particular energy subsidy is insufficient to increase brain size or production rate.[31]

The fossil record indicates that during the early million or so years after the emergence of mammals, there were many evolutionary experiments. Most species became extinct, probably for diverse reasons. Perhaps the litters were too big or the brains too large relative to the heart and lungs, or perhaps any one of thousands of flaws brought the species down, given their ecological conditions. Despite all the extinctions, mammals and birds have been highly successful evolutionary innovations. At this point, we know of roughly fifty-four hundred species of mammals, and about eighteen thousand species of birds. Among species, variability in how they manage their social lives reminds us that ecological conditions shape what evolution favors.

In mammals and birds, attachments to mothers, and in some cases to fathers, kin, and friends, is the platform for social behavior in general, and for moral behavior in particular. That basic platform works regardless of bodily configuration—for whales that live entirely in water or monkeys that live entirely on land. The adaptations to the environment are wonderfully diverse, and each species has a unique complement of social styles suited to surviving well enough to pass its genes on. For example, marmosets and titi monkeys are cooperative breeders (males and females share parenting duties), and they mate for life, but not chimpanzees or bonobos. In wolf packs and meerkat troops, there is only one breeding pair, but not so for baboons. Typically, female chimpanzees leave the natal troop when they mature; in baboons, the males leave the natal troop. Many species, such as brown bears and orangutans, do not live in groups, but socialize only to the extent of mating and caring for offspring. And so the list goes on, with beguiling diversity.[32] The scaffolding on the basic attachment platform can change to accommodate different ecological pressures, creating species-typical attachment patterns.

WHY DO HUMANS HAVE SUCH A BIG CORTEX?

Mammalian sociality is qualitatively different from that seen in social animals that lack a cortex, such as bees, termites, and fish. It is more flexible, less reflexive, and more sensitive to contingencies in the environment and thus sensitive to evidence. It is sensitive to long-term as well as short-term considerations. The social brain of mammals enables them to navigate the social world, for knowing what others intend or expect. Body language has evolved to signal feelings and goals, and brains have evolved to interpret those signals. In some mammals, the brain makes possible the accumulation of knowledge across generations, as offspring are taught the discoveries of their forebears. This happens on a grand scale in humans but is also seen, to a lesser degree, in chimpanzees, capuchin monkeys, and in some bird species.[33] In mammals, the cortex has a lot to do with those behaviors.

The last common ancestor of hominins and chimpanzees lived about 5 to 8 million years ago. Chimpanzees have been evolving from that common ancestor for precisely as long as hominins have.[34] The hominin brain, especially the hominin cortex, expanded enormously. The chimpanzee brain has remained largely the same size as it was. The brain of *Homo sapiens* is about three times the size of the chimpanzee brain.

The big expansion of the brain in hominins had to be paid for, either by increased energy resources or by reduced energy expenditure. Learning to cook food over fire was quite likely the crucial behavioral change that allowed hominin brains to expand well beyond chimpanzee brains, and to expand rather quickly in evolutionary time. Chimpanzees do not use fire or cook. This link between brain size and the use of fire to cook has been proposed by the anthropologist Richard Wrangham[35] and endorsed by anatomist Suzana Herculano-Houzel.[36] The claim is derived from the data showing that cooked meat and cooked roots deliver more in nutrition and calories than does raw food.[37] Early hominins, most likely *Homo erectus*, began to use fire about 1.5 million years ago, long before *Homo sapiens* appeared, about 300,000 years ago. Using fire to cook may have been the way that hominins paid for the massive expansion of neuron numbers in their brains.

The expansion itself may have been allowed, rather than driven; that is, the expansion of neuron numbers may have been allowed by the extra calories provided by cooking, rather than driven by a pressing ecological need. It is possible that the genetic changes for increasing neuron numbers in cortex may frequently and rather easily occur, but unless the additional neurons can pay their caloric way, an animal carrying those genetic changes tends not to be as successful as other members of the same species are.[38] Being unusually smart relative to your fellow species members may do you little good if you cannot get enough calories to feed your brain.

According to the hypothesis, therefore, once cooking paid the energetic bill of added neurons, hominins that happened to have a larger brain could survive and reproduce. Given this neural luxury, hominins such as *Homo erectus* and *Homo neanderthalensis* may have started to use their expanded brains to do socially more complex things than just foraging for food all the livelong day. If they did not have to forage from dawn to dusk, as chimpanzees must, they had the leisure to do such things as tell stories and draw; they had the time to make boats, music, and fancier tools.

MUD, GLORIOUS MUD[39]

The suite of evolutionary adjustments that ultimately led to mammalian styles of sociality, including what we might call morality, was first and foremost about food. Altruism—incurring a cost to oneself to benefit another—emerged as a consequence of the need for mothers to care for their infants, which in turn was a response to the endothermy innovation. Energy constraints might not be stylish and philosophical, but they are as real as rain.

Do the humble origins of our conscience demean its value? My response is a resounding *no.* Commonly in biology, beautiful things emerge from rather ugly sources, as *Psilocybe cubensis* mushrooms emerge from cow patties, or butterflies from rather unsightly caterpillars. The dominant role of energy requirements in the ancient origin of human morality does not mean that decency and honesty must be cheapened. Nor does it mean that they are not real. These virtues remain entirely admirable and worthy to us social humans, regardless of their humble origins. They are an essential part of what makes us the humans we are.

Although this chapter has addressed the evolutionary origins of sociality, questions about mechanisms in the mammalian brain that enable complex social behavior, such as attachment to kin and friends, remained unanswered. Neuroscientists have made impressive progress in nailing down the details of the wiring and neurochemistry of attachment, whether between parents and offspring, or among mates, kin, or friends. These details yield further insight into what it is to have a conscience; to feel strongly motivated to cooperate; to defend, as well as to punish, those who intentionally cause grief. In the next chapter we will look a little more closely at the inside of conscience.

CHAPTER 2

Getting Attached

Attachment is a unifying principle that reaches from the biological depths of our being to its furthest spiritual reaches.

JOHN BOWLBY[1]

Attachments, whether to parents, siblings, friends, or lovers, are highly robust over time. They can also be complex and subtle, especially in animals with a very large cortex. Introspection reveals nothing about the neural substrate for affiliation and the yearning to belong, but only about how much our attachments mean to us. Affiliation and attachment are the kinds of mental phenomena that you might put on a list of hard neurobiological problems, alongside the substrates for consciousness and language learning. As Francis Crick, the codiscoverer of DNA, was fond of pointing out, however, the key to unraveling complexity at the macro biological level might well be a deceptively simple mechanism at the micro level.

Could there be, lurking down in the depths of the brain, a simple structural system, capable of effects that fan out into diverse interactions and hence diverse forms? Maybe. After all, DNA is a lot like that. To code, it uses merely four bases—A, T, G, C (adenine, thymine, guanine, cytosine)—diversely ordered in a sequence 3 billion bases long that is unique to each individual human. Every living thing depends on DNA, yet the forms of living things fan out into a breathtaking range.

One particular neurobiological discovery led me to suspect that for sociality, there might just be a relatively simple story in the offing. Larry Young, a neuroscientist working on hormones in the brain, was invited to the Salk Institute to give a talk, and while Young's announced topic—mate attachment in voles—seemed mildly interesting, it did not presage a blockbuster. We sat politely as he launched in by describing the mating behavior in different species of voles. Then came the dramatic neuroscience.

Prairie voles and montane voles seem, to a casual eye, to be very similar kinds of rodents. Nonetheless, there is a striking behavioral difference: After the first mating, male and female prairie voles are attached to each other for life.[2] By contrast, montane voles meet and mate, and then they go their separate ways.

The unusual social behavior of prairie voles can be observed in the wild, but to be clearer about what exactly pair bonding involves, Young and colleagues examined aspects of vole behavior in the laboratory. They noticed that the bonded prairie voles prefer to spend time close to each other; the montane voles prefer to be alone. Once bonded to his female, the male prairie vole will attack intruders into the nest, including other female voles. More generally, prairie voles like to be with others in a large community. Not so montane voles; they are loners. In both species, the female cares for the pups, but only in prairie voles does the male guard the nest and huddle over newborn pups to keep them warm and safe. If the pair is separated, the individuals mope, and their levels of stress hormone go up.

Although *monogamy* is not quite the right word, because even bonded individuals may partake of a little action elsewhere, prairie voles nevertheless spend most of their time with the bonded mate and their offspring. In that less-than-strict sense, long-term pair bonders such as prairie voles are called *socially monogamous*. And us? By and large, humans tend to be socially monogamous, at least for significant periods of time (serial monogamy), if not for life.

Social monogamy in prairie voles provoked Young and his colleagues to ask this question: What are the differences between the brains of prairie voles and montane voles that explain the striking difference in mate attachment?[3] They found an answer that was surprisingly simple. It turned on a pair of similar hormones, oxytocin and vasopressin. Compared to montane voles, prairie voles have a greater density of receptors for oxytocin, in one very specific part of the subcortical brain, the *nucleus accumbens* (Figure 2.1). In addi-

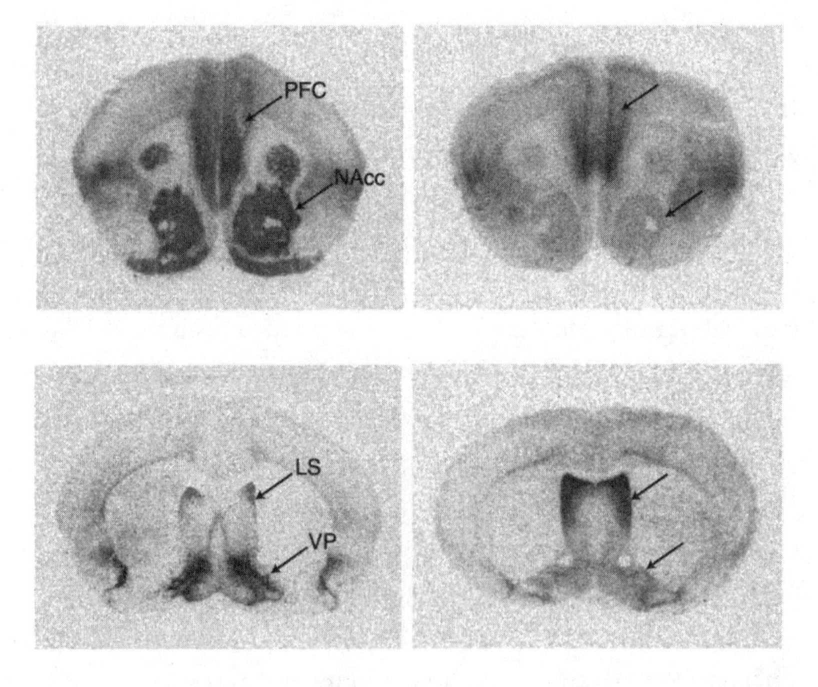

Figure 2.1 Slices of the brains of voles: prairie voles on the left, montane voles on the right. Oxytocin and vasopressin receptors have been stained so that they are visible. The image at top left shows a high density of receptors in the nucleus accumbens (NAcc). At bottom left, we see a high density of vasopressin receptors in the ventral pallidum (VP). The montane vole brains (right) are strikingly different in those areas. Interestingly, the montane vole has a higher density of vasopressin receptors in the lateral septum (LS), which may be associated with a preference for a nonsocial life. Both species have oxytocin receptors in the prefrontal cortex (PFC). LARRY J. YOUNG AND ZUOXIN WANG, "THE NEUROBIOLOGY OF PAIR BONDING," *NATURE NEUROSCIENCE* 7 (2004): 1048–54. WITH PERMISSION.

tion, male prairie voles have a very high density of receptors for vasopressin in an adjacent subcortical structure, the ventral pallidum. Montane voles do not. This was the dramatic answer. Incomplete, yes, but surprisingly simple: variation in the density of oxytocin and vasopressin receptors.

As background, bear in mind that a receptor is just a well-configured protein that sits on a neuron's membrane waiting for exactly the right neurochemical to drift along and fit itself into its slot. A neurochemical has no effect in the brain unless it can dock in its personalized receptor on a neuron's membrane (see Figure 1.1 in Chapter 1). Once the neurochemical is docked, the response patterns of the neuron may change, either boosting or reducing the likelihood that it will talk to other neurons, depending on the neurotransmitter. Basically, increasing the density of oxytocin receptors in a circuit tends to increase the clout of the oxytocin drifting around because more molecules will find a docking site. It is a little like increasing your taste buds (receptors) for sour things. If you have no sour taste buds at all, lemon juice will taste much like water. Increase the number of sour taste buds, and the sour taste of lemon juice is unambiguous. In a roughly similar way, increasing the numbers of neurons with receptors for oxytocin alters the circuitry in which those neurons work, which means it alters the behavior that the circuitry regulates.

Finding a correlation between monogamous pair bonding on the one hand, and receptor density for specific neurochemicals on the other, was remarkable enough, but to justify the stronger claim of *causality*, additional evidence had to be gathered. Various labs, including Young's, manipulated oxytocin and watched for behavioral effects in the voles. For example, they used a drug to block the oxytocin receptors in virgin prairie voles, then let them mate. Pair-bonding behavior was blocked. They injected oxytocin into the brains of virgin male and female prairie voles that were acquainted but had not mated. The voles acted as typical postcoital mates; they were "in love." Using genetic tools, the lab increased the vasopressin receptors in the ventral pallidum of male prairie voles, thereby enhancing the voles' mate preference by increasing their attachment behaviors, such as grooming and cuddling. Doing the same thing to montane voles caused them to behave like prairie voles—to prefer female montane voles with whom they had previously mated.

Moreover, Young and his colleagues went on to show that differences in oxytocin receptor density appeared to correlate with expression levels of a single gene that codes for the protein that is the oxytocin receptor.[4] The same was true for the vasopressin receptor density. These data reach beyond surface behavioral descriptions and into the explanatory depths. Now we are beginning to glimpse biological mechanisms.

After Young's talk ended, I wandered out of the room and sat on the Salk Institute's pool edge, looking out over the cliffs to the vastness of the Pacific. Francis Crick and I had often sat here and talked about the brain. Sometimes we had talked about morality and the brain. He once accompanied me to the University of California San Diego campus across the street to attend a philosophy seminar on ethics. As we walked back to the Salk, he expressed astonishment that the talk was all about pure reason, with nothing at all about the contribution of biology. Surely, he added in exasperation, philosophers must know about biological evolution.

Crick had thought it was highly likely that the basic motivation for sharing and cooperation, and for learning social norms, was fundamentally owed to the genes that build brain wiring. Until we got the biology nailed or at least semi-nailed down, Crick thought that focusing on reason was not getting to the heart of the matter. This seemed right to me. Much earlier, the Scottish philosopher David Hume (1711–1776) had argued that we are born with a predisposition to be socially sensitive—what he called our "moral sentiment." Crick's take on the matter seemed to be a modern version of Hume's eighteenth-century hypothesis, and his rationale was much like

Hume's: reason alone will never motivate typical moral behavior, even though it can help us figure out how to satisfy our moral desires.[5]

The trouble was that, at the time Crick and I had these conversations, there was no fruitful entry point from which to make progress in figuring out the brain-based nature of Hume's "moral sentiment." I could not see any neurobiological results that might spearhead a deeper understanding of moral behavior, or of our conscience. The prairie voles and their special receptor densities changed all that. Sadly, however, by the time Larry Young gave his talk, Crick himself had died.

Young's data regarding oxytocin and prairie voles were inspiring precisely because they *did* suggest an entry point into the brain-based nature of morality. His story made sense in neurobiological, psychological, and evolutionary terms. I was astonished to realize that a relatively tiny difference in structure—the density of receptors for oxytocin—could be at the root of something as apparently complex as monogamy. Equally astonishing was the fact that it is *oxytocin* that is at the core of mate attachment. Why? Because it is oxytocin that is at the core of mother-baby attachment. Could it be boiled down to this: *Attachment begets caring; caring begets conscience*? Modify receptor density in various regions with a small genetic tweak, and empathy may extend from offspring to mates, to kin, or perhaps to the wider community?

No faculty of reason governs vole monogamy; no religion lays down rules for the voles; no philosophical arguments are deployed at vole seminars. Prairie voles are socially monogamous because their neurobiology works that way. Individuals do not up and decide to be social on grounds that it might help them thrive. Our genes dispose us to be social mammals, and our thriving follows along. Evolution favored those developments. Moral norms emerge mostly as practical solutions to social problems, just as specific boatbuilding norms emerge as practical solutions to local water travel problems. Assuming that having a conscience involves caring for certain others with varying degrees of self-sacrifice, I could now see, albeit only in the most general terms, a path from biology to morality.

The evolutionary biologist E. O. Wilson had suggested in 1975 that the evolution of human sociality is the biggest conundrum facing biology. In 1975 he was probably right. By 2004, I was inclined to think that Wilson's biggest conundrum was beginning to fragment into a cluster of tractable, neurobiological puzzles. Species other than voles needed to be studied; the role of neurochemicals other than oxytocin and vasopressin needed to be understood; the role of cortical circuitry as well as subcortical circuitry needed to be investigated further. All that is granted. Nonetheless, it was profoundly encouraging that from the new vantage point, puzzles about our social nature looked to be empirical and experimental, not metaphysical and philosophical.

To risk repeating a point made in the Introduction, when we face a moral issue, the neurobiological data cannot tell us which is the morally preferred option regarding old-growth forests or capital punishment or insider trading. On the other hand, the data will help us understand why humans are generally motivated to care about those to whom we are attached and why social attachments matter so tremendously in life.

WHAT DOES OXYTOCIN DO?

By the time oxytocin was discovered to facilitate social attachment in mammals, it had long been known for its role in lactation (it is essential for milk ejection from mammary glands) and in uterine contraction during birth. It was, and still is, often used in humans to induce labor. Well before prairie voles became famous, the role of oxytocin in social behavior was anticipated by an experiment in 1979 in which oxytocin was directly injected into the brains of virgin female rats. A little later, oxytocin was directly injected into the brains of female sheep. In both cases, the injected females immediately began to exhibit full maternal behavior—something normally seen only in females who have just given birth.[6] For example, the treated females encouraged nearby young to suckle, and they began to lick the young, just as mothers do immediately after giving birth. These data clearly showed that oxytocin could stimulate complex social behavior.

The data on mate attachment in prairie voles inspired a range of experiments that generated results to round out the picture of the mechanism. Vasopressin, for example, was found to be more abundant in males than in females, and to be involved also in aggression, especially in the defense of infants and mates. In prairie vole communities, older siblings also help rear the pups, and unlike mice, they show strong incest avoidance. Prairie vole pups that were well fed but socially isolated grew up unable to form partner attachments. This observation tells us that postnatal experience has an effect on the circuitry for sociality.

Oxytocin also has a role in sensory processing, especially in the olfactory system. In rodents, smell has been shown to be important for recognizing offspring and intruders, and oxytocin affects sensory perception in the exploration and recognition of mates. Human mothers also recognize their babies' unique smell.[7]

Other socially monogamous species have been studied, including titi monkeys, owl monkeys, and marmosets. Wider patterns of oxytocin receptor distribution were found in socially monogamous monkeys than are seen in rodents, and the effects of oxytocin on social behavior are correspondingly more complex. One particularly striking result is that, in strongly bonded marmosets, fluctuations in oxytocin levels are synchronized between mates.[8]

In one particular experiment relevant to empathy, one partner of a pair of prairie voles is exposed to a stressor (such as movement restriction), and then returned to the cage to re-join the mate. Immediately, the unstressed partner hurries to the stressed partner and engages in intense consolation behavior—grooming and licking. Here is the control: if the pair of prairie voles are merely separated but no stress is experienced by the absent partner, a warm but less intense reunion occurs. If oxytocin receptors are experimentally blocked in the home-cage mate, intense consolation behavior toward a stressed partner is not seen.[9]

What is remarkable is that once the stressed partner is returned to the home cage, the levels of stress hormones in the home-cage partner increase to match those of the stressed partner. This observation suggests a mechanism for empathy, indicating that if one vole becomes highly anxious, the partner vole's brain reads the signs and responds so as to match that emotional state.[10] Casual observation suggests that this is very likely to be true of affiliated humans as well. Empathy, in some form, may well be typical of highly social mammals in general.

Among researchers, there were also the inevitable self-promoters, who hopped on the oxytocin bandwagon with rather flimsy knowledge of neuroendocrinology but sporting charm and marketable sound bites. Oxytocin was hailed as the "love molecule," the "moral molecule," and the "cuddle molecule," all names embellishing the data in misleading ways. Like snake oil, oxytocin was advertised as the cure for a range of ills, including social awkwardness, bad behavior at school, obesity, indifference in mates, and congressional inaction on social policy.

An important thing to remember is that oxytocin and vasopressin, though crucial in the brain's social network, are only two elements in a suite of neurochemicals that act on neurons. The neurochemicals in the suite also interact with one another, facilitating, or in some instances inhibiting, their actions. In the background are various hormones. Estrogen, for example, is coexpressed with oxytocin, and the pair act together to reduce stress.[11]

Recently, another neurochemical has been found to be important in parenting behavior: galanin, released from a group of neurons in a tiny region of the hypothalamus,[12] controls individual differences in pup care. Shut down galanin neurons, and mothering in lactating females fades, and they are prone to ignore their pups. Male behavior can be affected too. A male mouse tends to kill baby mice, until his own litter is about to be born. Then his good-dad behavior kicks in. What is the difference in his brain? Galanin. Here is some evidence: If the neurons releasing galanin are ablated, his good-parenting behavior stops. Moreover, normal infanticidal male mice will cease killing if their galanin-producing neurons are artificially stimulated.[13]

A peptide is a molecule that consists of a string of amino acids. Oxytocin and vasopressin are simple peptides, each having only nine amino acids. The lineage of oxytocin and vasopressin goes back at least 500 million years, long before mammals first began to appear. The vasopressin and oxytocin seen in mammals probably evolved from a single peptide, perhaps vasotocin, found in amphibians and reptiles, or isotocin, found in fish. A slightly different variant, nematocin, is found in the tiny worm *Caenorhabditis elegans*, whose entire nervous system has merely 302 neurons (human brains have 86 billion neurons).

What does this homolog of oxytocin do for the worms? Surprisingly, when *C. elegans* larvae begin to populate a patch, they release nematocin, which binds to receptors in the adult worms, causing the adults to leave the patch to feed elsewhere, thus letting the larvae feast without adult competition. To a first approximation, this behavioral effect seems like a simple bit of parental sacrifice for the offspring.[14] Do the adult worms feel affection for their sweet little larvae? A mere 302 neurons are almost certainly insufficient for that. In this case, genes see to it that the sacrifice is made, and no moral agonizing is involved.

In reptiles and fish, oxytocin homologs play various roles in fluid regulation and in reproductive processes such as egg laying, sperm ejection, and spawning stimulation. In addition to its role in the regulation of social behavior, oxytocin is important for various bodily functions related to mating. In mammalian males, oxytocin is secreted in the testes and is necessary for sperm ejection. In mammalian females, it is secreted in the ovaries and plays a role in the release of eggs.[15] It is also found in the heart and in the gut. Vasopressin in mammals is important for maintaining the right water balance in the body. To prevent dehydration, it will reduce urine output by stimulating water reabsorption by the kidneys.

Although these facts seem a long distance from the topic of what it is to have a moral conscience, I like to savor them because they remind us that evolution is not an engineer that designs a device from scratch, but a blind process that, without any goal, fiddles around with the structure already in place. Outcomes of the fiddling are frequently less than optimal from an engineering standpoint, but as long as they work well enough to give the animal an edge in the struggle to survive and reproduce, they get passed on. In the evolution of the mammalian brain, oxytocin was repurposed to serve a specifically social function.

During pregnancy, genes in the fetus and in the placenta make hormones that are released into the mother's blood (e.g., progesterone, prolactin, and estrogen; Figure 2.2). The release of these hormones leads to a sequestering of oxytocin in neurons in the mother's hypothalamus. Just before delivery, the density of oxytocin receptors in the hypothalamus surges.[16] Vaginal-cervical stimulation—the normal effect of giving birth—causes a flood of oxytocin to be released from the hypothalamus into other parts of the brain.[17]

The hypothalamus is a small, ancient structure in the brain whose components are essential for many basic life functions, including feeding, drinking, aggression, and sexual behavior (Figure 2.3). In mammals, the hypothalamus secretes oxytocin to particular brain locations, thereby triggering a cascade of events with the end result that the mother behaves maternally and becomes powerfully attached to her offspring.[18] The hypothalamus also secretes vaso-

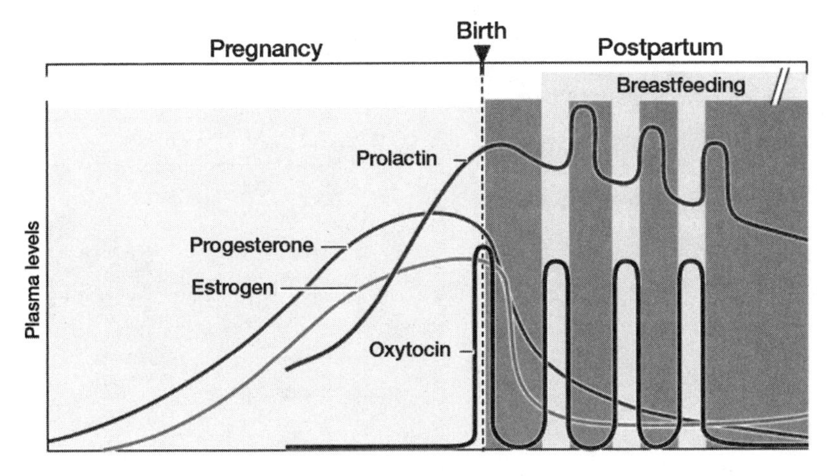

Figure 2.2 Hormone levels during and after human pregnancy. Estrogen, progesterone, and prolactin steadily increase throughout pregnancy. Childbirth is characterized by a rapid drop in estrogen and progesterone and a surge in oxytocin levels that initiates uterine contractions. During breastfeeding in the postpartum period, pulses of prolactin levels stimulate milk production between feedings, alternating with oxytocin pulses, which lead to milk ejection during breastfeeding in response to the infant's suckling (the letdown reflex). JOHANNES KOHL, ANITA E. AUTRY, AND CATHERINE DULAC, "THE NEUROBIOLOGY OF PARENTING: A NEURAL CIRCUIT PERSPECTIVE," *BIOESSAYS* 39, NO. 1 (2017): 1–11. WITH PERMISSION.

pressin, which triggers a different cascade of events, motivating the mother to protect her offspring, including defending them against predators. Oxytocin-expressing neurons project their axons to the amygdala (see Figure 2.3), which plays a role in generating emotions including fear but also joy.[19] At the target sites, the neurons release their oxytocin. One effect of releasing oxytocin in the amygdala is that fear is dampened. This is probably why it helps to cuddle a child who wakes from a nightmare. The cuddling releases oxytocin, calming the child down, reducing anxiety and fear.

These oxytocin-expressing neurons also send their axons to a range of other brain areas, which include ancient parts of the reward system, as well as to the cortex, especially to the orbitofrontal cortex (the bit of cortex just above your eyeball orbits; see Figure 2.3). In mammals, one such area in the reward system with receptors for oxytocin is the nucleus accumbens, known to play a crucial role in

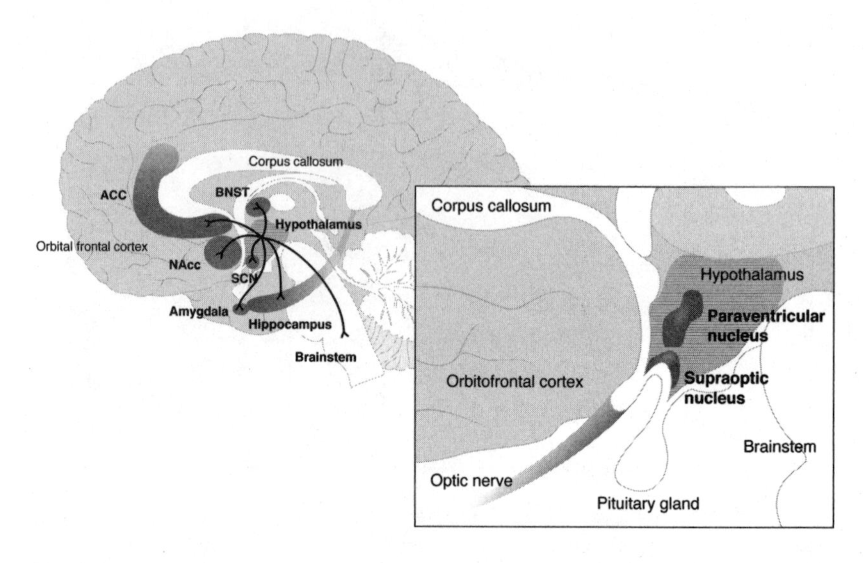

Figure 2.3 Left: Diagram of the right hemisphere of the human brain, seen from the midline, as though the left hemisphere were removed. Black lines depict selected neuronal pathways from oxytocin-releasing neurons in the hypothalamus to the anterior cingulate cortex (ACC) and to subcortical structures such as the nucleus accumbens (NAcc), the amygdala, the suprachiasmatic nucleus (SCN), the bed nucleus of the stria terminalis (BNST), and the brainstem. Right: Magnified diagram of the hypothalamus and its two oxytocin-releasing structures: the paraventricular nucleus and the supraoptic nucleus. Other nuclei in the hypothalamus, such as those regulating eating and drinking, are not shown. ADAPTED FROM A. MEYER-LINDENBERG ET AL., "OXYTOCIN AND VASOPRESSIN IN THE HUMAN BRAIN: SOCIAL PEPTIDES FOR TRANSLATIONAL MEDICINE," *NATURE REVIEWS NEUROSCIENCE* 12 (2011): 524–38. WITH PERMISSION.

wanting (being motivated to seek something) and in liking (getting pleasure from the object or event). As shown in Figure 2.1, the density of such receptors in this region is much higher in monogamous voles than in promiscuous voles. What does this difference signify? It tells us that many neurons in monogamous animals change their activity levels when oxytocin binds to their plentiful receptors. Those neurons are part of the circuitry that supports caring for mates.

Activation of one group of specialized neurons in the nucleus accumbens is also necessary for the mother to recognize her own pups; activation of another group triggers the motivation to engage in maternal behavior. The nucleus accumbens has links to the posterior region of the ventral pallidum, a kind of gateway in the reward system, as well as a motivational "governor." The ventral pallidum has an opioid hot spot (lots of opioid receptors) where stimulation increases liking and wanting; on its front end is an opioid cold spot, where there are fewer opioid receptors, and where stimulation decreases liking and wanting.[20]

As noted in Chapter 1, cannabinoids are marijuana-like neurochemicals made by the brain. When cannabinoids dock on specialized receptors, we get positive feelings.[21] They are released in the nucleus accumbens, in response to an oxytocin hit. Cannabinoid receptors exist also in the ventral pallidum, likely playing a role in a pleasurable outcome of an action. Binding of the cannabinoids to their receptors is important for the rewarding aspects of various kinds of social interactions, such as parenting and pair bonding.[22]

The endogenous (brain-made) opioids may be released when highly social animals are with kin or friends. One effect is to dampen the pain response. Or put another way, the threshold for pain is raised when animals are socializing, compared to when they are isolated.[23] The endogenous cannabinoids and opioids, along with their portfolios of receptors, are a major source of the pleasure we take in our social lives. The pleasure we get from cuddling our mates or our babies is an internal reward signal that reinforces the behavior. More cuddling ensues. More opioids are released.[24]

Oxytocin turns out to be crucially involved in the recognition of one's own offspring, as well as of one's mate, kin, or friends. The recognition circuitry seems to be the same for mating adult voles as for mother-infant recognition, and the motivational circuitry seems basically the same as well. It requires oxytocin and vasopressin, as well as the ventral pallidum, and the reward system more generally.[25] Incidentally, when rodent fathers, such as prairie voles, do care for their young, the brain circuitry activated seems to pretty much match that seen in females. In species where the male mates but then moves on, as is typical with montane voles, the social circuitry does exist in males for parenting and for mate preference, but it is inhibited by the absence of a suitable level of receptors for oxytocin and vasopressin, or perhaps also by those same receptors located in a different part of the circuitry.[26] As noted earlier, a relatively small genetic tweak, such as increasing the density of vasopressin receptors in the ventral pallidum, is all it takes to turn a promiscuous vole into a monogamous vole.[27]

One striking effect of oxytocin is that it decreases the stress response. To a first approximation, when oxytocin levels in the brain rise, the level of stress hormones falls. Thus, stress and anxiety can be buffered in prairie voles by the close presence of a mate or by the administration of oxytocin directly into the brain.[28] Similar anxiolytic (anxiety-reducing) effects of mates, kin, and friends, presumably involving oxytocin release from the hypothalamus, are seen in humans. This observation fits well with people's general social understanding that someone who is grieving or otherwise troubled is apt to benefit from the social support of friends and kin.[29]

To explain more precisely how oxytocin achieves its bonding effects, Larry Young and his colleagues have suggested that a major effect of oxytocin and its receptor portfolio is to narrow the rewarding effects of a behavior in such a way that each vole involved in the mating wants and likes *only* that one partner and no other. After the first mating, the male prairie vole associates the rewarding experience with exactly one smell. He therefore seeks exactly one female to recreate and repeat his pleasurable experience. Rather than "playing the field," as montane voles do, he will spend lots of time cuddling with his preferred mate, and will ignore or even rough up other females he encounters. (The same is true, with the obvious adjustment, for females.) In other words, perceptual attention is narrowed to the preferred partner. "I only have eyes for you," as the song goes. Or, if you are a vole, only you smell right for me. Oxytocin fixates attention and, therewith, perception and behavior.

Undoubtedly, montane voles find their mating experience equally rewarding, but according to the hypothesis, they do not associate it with one specific female, because they lack the oxytocin and vasopressin receptors in the reward system. Instead, the male montane vole goes off in search of any female whose odor indicates she is in estrus. He is not selective.

This attentional-focus hypothesis is compelling, but it is not the whole story. Monogamous species, such as prairie voles, beavers, titi monkeys, and many humans, bond for life. Because the rapture of early love is apt to wear off after some months or years, the strong attachment seen in the long term may require some other changes. Being able to trust and count on a partner during life's adventures is a real value and suggests further changes in the brains of lifelong partners to maximize this value. The brain may build on the early rewarding and focused perceptual experiences to establish a single favored partner, but in the long run, the bond may be maintained by a somewhat different set of operations in the reward system. Habits of many kinds make life easier, more predictable, and energetically favored. The one-and-only habit may have many advantages, and may be reinforced by frequent grooming, licking, and cooperating. I do not mean *habit* in a tacky way. Rather, I have in mind the kind of habit, such as making nutritious meals or jointly building a home, that contributes to a good and satisfying life.

This hypothesis is plausible. Among other things, it helps to explain why the disappearance of a mate causes the remaining partner to show loss behavior, including the typical signs of animal sadness—namely, loss of appetite, lethargy, and stress.

Since the reward system plays an important role in maintaining mate bonding, we might wonder whether it also plays a role in enduring friendships in groups where mating is not followed by pair bonding. Among friends, affective touch (hugging, stroking), consoling, sharing food, and defending against within-group aggressors raise levels of oxytocin and lower levels of stress hormones. Anxiety is therefore reduced, and calm increased. The reward system responds accordingly.

The role of experience in the world adds to the complexity behind the story told so far. For example, just *watching* social interactions can itself change the brain. Let's look at an example. *Nulliparous* (meaning "has not yet had babies") female rodents are apt to kill or ignore nearby pups. Over a few days, this behavior wanes if the nulliparous females are continuously exposed to lactating mothers and their care of their pups. Under these conditions, such stimuli change the nulliparous rodent's brain, weakening her urge to kill and instead fueling her maternal instincts. Nulliparous female rats thus exposed will retrieve pups into a small location and may even try to get nearby pups to suckle, a behavior perhaps reminiscent of children who use their dolls to mimic mothers nursing their babies.

This effect from merely watching indicates that the oxytocin-related circuits involved in the motivation to mother are sensitive not only to events that happen to the female rat, such as parturition (giving birth and hence vaginal-cervical stimulation), but also to what she observes around her, such as a dedicated mother caring for her pups. Not unrelated, sometimes in monogamous species the exposure of males and female to each other without an actual mating event triggers strong partner preference. Various sensory stimuli, such as smell, tactile and visual cues, and perhaps even auditory cues, make clear which partner is favored.

To summarize, the basic story is that in highly social mammals, oxytocin is released in the brain in positive social situations, such as grooming, cuddling, sex, and food sharing.[30] At least in highly social animals, this release, supported by the cannabinoids, tends to intensify social attachment. It results in reduced vigilance and anxiety, and an increased sense of trust and well-being. This outcome is rewarding, and indeed, the reward system responds to reinforce the behavioral routine. In this favorable state, trust and cooperation is facilitated, in turn strengthening the bonds between us. In this manner do we social mammals become even more disposed to care and share, and to adhere to social norms that produce approval from those we love.[31]

MY ONE AND ONLY

Although essentially all mammals provide maternal care to offspring, only about 5% have lifelong partners. Some are rodents, such as prairie voles and California deer mice, and some are primates, such as gibbons and New World monkeys like titis, marmosets, and capuchins. And then there are wolves, which have a complicated social life. In the pack, only the alpha female and the alpha male breed, and they continue this pattern until one dies. They are visibly affectionate with each other.[32] All in the pack help with the pups, and the alpha male helps his mate clean out the old nest in preparation for the next litter of pups. Beavers mate for life and share the responsibilities of parenting. Zoologists tell us, however, that the great majority of mammalian species (95%) are not monogamous and not biparental in the care of their young. By contrast, the great majority of bird species are.

Why, adaptively speaking, is monogamy the preferred mate arrangement in any species? Surely, promiscuity in males would always be advantageous. Several answers have been proposed. One is rather obvious—namely, that when ecological conditions are especially unforgiving, two parents sharing the care of the young are more reproductively successful than is a single parent carrying the whole burden. Prairie voles live on the open prairie and are subject to predation that is more intense compared to that of the more sheltered montane voles. Biparental systems are obviously preferable in most bird species. If one parent leaves the nest unattended to forage for food, she is likely to come back to find that a hawk or kestrel has picked off the babies as food for its own brood.

An additional possibility suggested by biologists is that biparental arrangements curb male infanticide. Male brown bears and polar bears, for example, regularly try to kill bear cubs if they suspect (probably by smell) that they were sired by another bear. If the male is successful, the female immediately goes into estrus, and the infanticidal male has a chance to father her next cubs. When males are biparental, however, as in prairie voles and beavers, infanticide of their own pups is essentially nonexistent.

ARE WE LIKE PRAIRIE VOLES?

Data from anthropologists and psychologists indicate that humans generally seem to be inclined to long-term bonding: even if they do not bond with one person for life, they tend to have strong partnerships that last many years. In many countries in the world (e.g., Japan, China, the US, Canada), polygamy is not legal, though of course sexual activities outside of marriage do sometimes occur. The nature of the wider ecology, and the availability of resources for food and shelter, are likely to have a significant impact on marriage conventions. Among hunter-gatherer-scavengers, monogamous arrangements are common, but polygamy (many wives for one man) is customary in some religions, and in a very few cultures polyandry (many male sex partners for each woman) is customary. Where conditions are very harsh or where, for practical reasons, it is difficult for men to support more than one wife, monogamy tends to be the dominant marital arrangement.[33]

Among the Inuit people of Baffin Island in the Arctic, when first visited by the anthropologist Franz Boas in 1883, marriage practices were informal, and long-term bonding, typical. Sharing a wife for a year or so was acceptable but infrequent, and spunky wives would just up and leave a husband who treated them badly. As Boas observed more generally, the Inuit did not have fixed rules about most aspects of social life, but seemed to acquiesce in the wisdom of past practices to loosely govern matters such as marriage and divorce. Boas was intrigued to note that the Inuit groups he encountered were rather anarchic by prevailing European standards, and he marveled at how well they managed in such harsh conditions without strict rules but with great reliance on historical tradition.

To determine whether our attachments to mates and friends share features of the mechanism found in prairie voles, experiments relevantly similar to those for voles would be methodologically ideal. Administering oxytocin or an oxytocin blocker directly into specific brain regions of voles and monkeys has been singularly successful in producing meaningful data about the role of oxytocin in the animal's social behavior. Ethical reasons, however, rule out these kinds of interventions as not suitable for humans. To find an ethical method for altering oxytocin levels in the brain before measuring behavior, researchers hit upon the idea of putting oxytocin in a nasal spray with the aim of getting it into the brain. Fortunately, spraying oxytocin up the nose is safe, painless, and quick. Subjects do not mind a bit.

The first report of results using this strategy was published in 2005 by Michael Kosfeld and his colleagues.[34] They wanted to test whether elevating levels of brain oxytocin would increase trusting behavior. Their subjects played a two-person investment game in which trusting the anonymous partner involves some monetary risk, but if trust is maintained across exchanges, both partners make real money. Twenty-nine subjects got an oxytocin nasal spray before the game began; twenty-nine control subjects got just a saline spray. In this sample, the results were intriguing: those who received oxytocin in their nasal spray showed greater levels of trust in playing the game and hence made more money than did the control subjects.

The Kosfeld results sparked many experiments in which oxytocin in a nasal spray was administered to test its effects on socially relevant capacities such as recognition of facial expressions (sad, angry, happy, and so forth), or the willingness to help or to trust, and on levels of affection for a mate. In many reports, the results showed that nasally sprayed oxytocin did enhance social capacities in humans, often as predicted.

Then, neuroscientists raised inconvenient questions. One question was, How exactly is oxytocin supposed to get from the nose into the brain?[35] Given their positive findings, various labs did not worry overmuch about that question. If it works, who cares? Cocaine gets into the brain when snorted, so why not oxytocin?

There is an answer, and unfortunately the answer complicates research life. Between the brain and the vascular system is a complex membrane. It constitutes what is known as the *blood-brain barrier.* Functionally, the blood-brain barrier protects the brain against infections and toxins. Some chemicals, and cocaine is one, do cross the barrier quite easily; some are totally shut out; some cross only with difficulty. Oxytocin is known to be in the "with difficulty" class. Oxytocin in a nasal spray probably does not readily cross the blood-brain barrier to reach the brain.

Those who used the nasal-spray technique replied that if the experiments have positive results, then somehow, some way, nasally sprayed oxytocin must get into the subject's brain, and in sufficient quantity to have a behavioral effect. Although their point is not unreasonable, scientific caution bids us look closely at the experimental details behind those positive results.

Two lines of questions are pressing: (1) Do the intranasal experiments have the statistical power to justify meaningful conclusions about the effects of nasally sprayed oxytocin in the human brain? (2) How could oxytocin get into the brain if sprayed up the nose?

On the statistical-power question, a meta-analysis of intranasal oxytocin experiments revealed that many studies are seriously underpowered or suffer from methodological flaws. Wallum, Waldman, and Young put the results of their recent meta-analysis bluntly:

> Our conclusion is that intranasal oxytocin studies are generally underpowered and that there is a high probability that most of the published intranasal oxytocin findings do not represent true effects. Thus, the remarkable reports that intranasal oxytocin influences a large number of human social behaviors should be viewed with healthy skepticism, and we make recommendations to improve the reliability of human oxytocin studies in the future.[36]

The worry, therefore, is that in some cases, what looks like a positive result may be a statistical artifact. Even in the original Kosfeld experiment, the statistics reveal that the effect of oxytocin administration is actually very small, accounting for only 17% of the variance. Among the published results, it is now unknown which of them contain false positives. Future experiments can solve the problem of statistical power, and we will have to wait and see what the results of better-designed experiments show. Incidentally, reports from single individuals who ordered online nasal spray containing oxytocin are unreliable because of the placebo effect and also because the contents of such nasal sprays are not regulated and hence could be just about anything.

As for the blood-brain barrier, the unanswered question is whether some oxytocin may leak through the barrier into the brain. So far, research has not shown such a leaky path into the brain, or, more exactly, into regions of the brain where there are oxytocin receptors for the oxytocin to bind to. That matter urgently needs to be clarified, if the results from the intranasal method are to be trusted.[37]

Because social impairments are seen in subjects with autism, it occurred to researchers that intranasal oxytocin interventions may be a promising treatment for individuals with this condition. Although early research showing modest positive effects stirred hope, regrettably the results failed to be replicated in appropriate experiments.[38]

In addition, the hypothesis that those on the autism spectrum—that is, those with autism spectrum disorder, or ASD—have a deficit in oxytocin levels or in oxytocin receptors has not been supported in genetic tests.[39] Although genetic variants for oxytocin or its receptors do correlate with social impairments, this correlation is neither specific to those with ASD nor typical among them. The suggestion, then, is that a cause more basic than social impairment is responsible for autism. To put it another way, social impairments like those we see in autistic subjects are probably owed to abnormalities other than genetic variants that specifically alter oxytocin and its receptors.

Because the density of oxytocin receptors in the nucleus accumbens is important in regulating social behavior in prairie voles, an intriguing question regarding humans is what the density of oxytocin receptors in our nucleus accumbens looks like, and how much variability there is across populations. How can we answer that question?

At this stage, receptors can be identified only in postmortem tissue, not in living tissue. The technique for locating receptors involves injecting a radioactive label into the brain. The label is designed to attach to only the protein being sought, such as the oxytocin receptor. Postmortem, the brain tissue is thinly sliced to reveal the location of the radioactive label and hence the receptor. One recent study tested for oxytocin receptors in two deceased human females.[40] The researchers did find receptors in the nucleus accumbens, amygdala, and hypothalamus, comparable in density to what has been seen in prairie voles and titi monkeys. So far, so good. A sample of only two brains, however, means that no strong conclusions can yet be drawn about humans.

Some experiments have tried to determine whether levels of oxytocin in the human brain change owing to a specific social interaction. Do they go up after a nice massage or a hug, for example, or down if you are shunned by others at a party? For ethical reasons, experimenters cannot directly take fluid from the brains of humans, since that requires a spinal tap—a procedure that must be performed by a physician, carries significant risk, and causes post-tap headaches. Is there an indirect measure of oxytocin in the brain? How about measuring changes in oxytocin levels in the blood? Blood samples, after all, are easy to get.

Appealing though the idea is, the problem is that the pathway by which oxytocin enters the body is different from the pathway for its release into the brain, and the two forms of release appear to be uncoordinated, as far as we know. Consequently, blood levels of oxytocin may not tell us much about brain levels of oxytocin. Levels of oxytocin can also be measured in urine and saliva. Here again, it is not known how accurately these measurements reflect oxytocin levels in those places in the brain where oxytocin does its work. Brain and urine levels of oxytocin may, or may not, strongly correlate. This methodological problem will likely be solved soon.[41]

Disappointing as the cautionary flags are, we have to acknowledge that the data are only as good as the methods deployed. Enthusiastic descriptions do not make the data any better than they are. Patience eventually pays off, and it is probably wiser to be a bit of a nitpicker than a gullible investor.

Although many researchers would dearly love to figure out ethically acceptable ways of obtaining reliable data at the neural level on humans and their oxytocin portfolio, at this stage we rely largely on inferences from monkeys, rodents, and other mammals. While this strategy may not be ideal, it is productive as long as we recognize that our inferences depend on what we know about shared brain organization.

ELEMENTS OF CONSCIENCE

The neural wiring for attachment and bonding provides the motivational and emotional platform for sociality, which enables a scaffolding of social practices, moral inhibitions, and norms. If mammals did not feel the powerful need to belong and be included, if they did not care about the well-being of kith and kin, then moral responsibility would have no toehold.

Interlaced with the platform for other-care, learning mechanisms respond to experience to build a complex brain model of the social world, shot through with emotions, values, and social practices. This inner model enables us to recognize what others are feeling and intending, and to get along in the social world. When animals are attached to each other, they are less apprehensive and more trusting. Consequently, cooperation, grooming, food sharing, mutual defense, and so forth are more apt to occur when trust prevails.

Bonding to others in our social world, and hence caring what happens to them, is a profoundly significant feature of our nature as human animals. Importantly, however, social affiliation coexists with self-care. We do not cease to care about ourselves just because we are also bonded to others. We all contend with the fact that self-care and other-care typically abide together in what is often a delicate balance. Too much of one, and we are scolded for being selfish. Too much of the other, and we are reproached for neglecting ourselves in pursuit of imprudent do-gooding.

Acquired patterns of caring behavior—habits and norms—take shape during development as we learn how to behave with others. The reward system internalizes social norms by using imitation along with the pleasure of social approval and the pain of social disapproval. We come to feel unpleasantly anxious when we are tempted to steal or lie; we feel anticipatory satisfaction when we plan to soothe a hurt friend or help with a new baby. More sophisticated social behavior develops over time. The cortex supports flexibility and intelligence in the means by which we express how we care.

Between them, the circuitry supporting sociality and self-care, and the circuitry for internalizing social norms, create what we call *conscience*. In this sense your conscience is a brain construct, whereby your instincts for caring, for self and others, are channeled into specific behaviors through development, imitation, and learning. In the next chapter we will look at the neural substrate for learning norms and how it interlocks with the platform for sociality.

CHAPTER 3

Learning and Getting Along

By three methods we may learn wisdom: First, by reflection, which is noblest; second, by imitation, which is easiest; and third by experience, which is the bitterest.

CONFUCIUS, *ANALECTS*

Confucius's observations hold for learning the ways of the social world, as well as the ways of the physical world. They hold for the shaping of a conscience from infancy through to the end of life, as well as for the shaping of a golf swing or a surgical technique. Notice, though, that in his comment, Confucius omits the *positive* aspects of learning from experience. His omission probably reflects a common feature of our autobiographical memories. It is often the embarrassments and faux pas, the missteps and blunders, that we selectively recall as we contemplate what we have learned through experience. Positive reward is, however, an essential aspect of learning from experience, and especially for learning the norms and practices of our social milieu. Social approval, inclusion in the group, and shared laughter are all highly rewarding.[1]

Training through reward is strongly linked to feelings of anxiety or satisfaction. A dog, for example, will show discomfort if asked to violate a restriction it has been trained to abide by. I learned this when quite young. Our farm dog, Nick, was trained not to venture into the house beyond the kitchen. When he was about three years old and I was home alone with not much to do, I decided to see whether I could get him to cross the line into the living room. I called to him from just beyond the boundary line. I reassured him that it was okay just this once. He stared at me, his tail went down, his head lowered. He was conflicted. But he would not budge, call as I might. I upped the game by holding out a piece of sausage. He looked embarrassed and hung his head, but he did not come forward. He backed up, turned, and walked out of the house. Now it was my turn to feel ashamed for having asked Nick to do something he had been trained was wrong.

Although it has long been known that reward training is a powerful technique for shaping behavior, the mechanisms that enable such effects have been hidden in the neural circuitry until recently. How does training overcome instincts and strong desires? How can training result in a behavior that was never produced in our evolutionary condition, such as driving a car?

One of the electrifying neuroscience stories of the last three decades has been the methodical discovery of the mechanisms supporting *reward learning*—also known as *reinforcement learning*. The same basic mechanisms appear to be deployed in acquiring both social and nonsocial norms—in how to behave if serving on a jury, in how to change a tire, in how to lend a hand in troubled situations, in how to right a flipped canoe. Different memories and background skills come into play, but the same reward mechanisms are at work.

At the heart of the story of reinforcement learning in mammals are the homologs of neural structures that arose long before the emergence of the mammalian brain or even the reptilian brain. Primordial circuitry, with its ancient organization and its ancient way of operating, is foundational.[2] It controls the basic functions of survival, including feeding, mating, and avoiding predators. It enables learning where to find reliable food sources, and when to change foraging patterns. In mammals, the circuitry is found in the midbrain, coordinated with a syndicate of structures called the *basal ganglia*[3] (Figure 3.1). The cortex, especially its frontal regions, interlaces with the basal ganglia, extending and modifying its range of action and thus enabling high-level control.

Richly connected to the cortex, the hippocampal structures support ongoing memory for specific events, and for the characteristics of individuals—that Uncle Hamish is irritable, that Auntie Martha

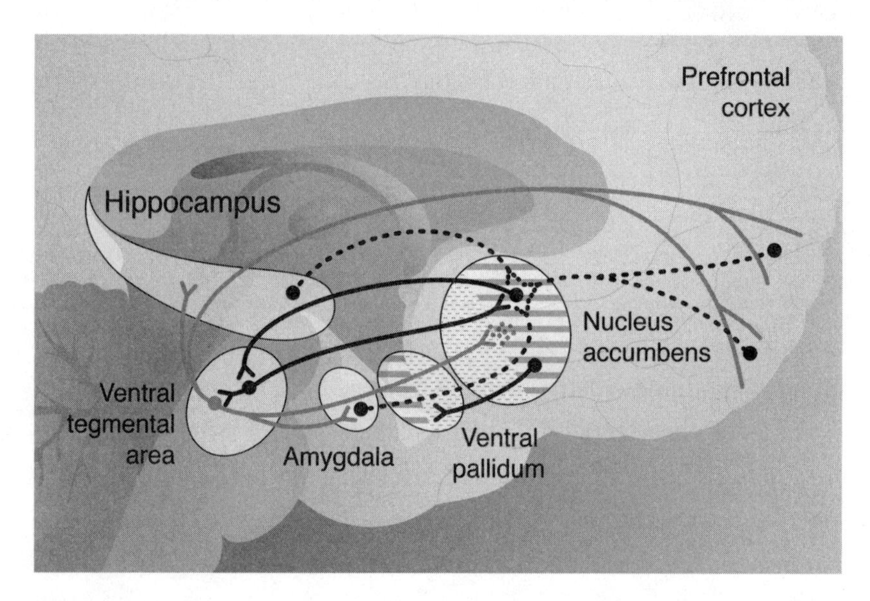

Figure 3.1 Simplified diagram showing the major components and connections of the reward system in the human brain, as seen from the midline of the brain, with only one hemisphere (the left) visible. Connections among subcortical nuclei and between those nuclei and the prefrontal cortex are very rich, but they are depicted here by only a single neuron for each major pathway. The axon terminals of neurons are signified by a V shape. Pathways differ according to the neurochemicals released: those represented by a dashed line release glutamate, an excitatory neurotransmitter; those represented by a solid black line release GABA (gamma-aminobutyric acid), an inhibitory neurotransmitter; those represented by a solid gray line release dopamine, a neuromodulator critical for reinforcement learning. The regions of the nucleus accumbens and the ventral pallidum with horizontal gray lines are hedonic hot spots containing opioid and cannabinoid receptors, where stimulation enhances liking reactions; the stippled regions are hedonic cold spots where liking reactions are suppressed. The hypothalamus is not drawn, because it would be obscured by the ventral pallidum. INFORMATION ABOUT HOT SPOTS AND COLD SPOTS IN THE NUCLEUS ACCUMBENS AND VENTRAL PALLIDUM WAS DRAWN FROM D. C. CASTRO AND K. C. BERRIDGE, "ADVANCES IN THE NEUROBIOLOGICAL BASES FOR FOOD 'LIKING' VERSUS 'WANTING,'" *PHYSIOLOGY OF BEHAVIOR* 136 (2014): 22–30.

tells off-color jokes. Via the hippocampus, daily experiences that are deemed worthy of recollection are integrated into our stored background knowledge, thus extending the capacity for skillful navigation of the world.

The bigger the cortex relative to the basal ganglia, the more far-reaching and intricate the learning by experience. The bigger the cortex, the greater the capacity to acquire abstract models of how the world works, to draw upon those models when appropriate, and to use feedback to update those models.

In big-brained mammals, including us, a goal devised now may not be achieved until quite far into the future. The means of achieving that goal may involve a host of intermediate steps.[4] Consider the steps involved in building a house or removing an inflamed appendix. Precise sequencing of steps is essential for reaching many goals, especially those that involve other smart animals. Local contingencies may shape the exact nature of those steps, as we figure out what to do when something unanticipated suddenly occurs.

Humans are not the only mammals to devise multistep plans. Consider the cunning decisions made by a grizzly mother as she succeeds in bringing down a caribou to feed her hungry cubs.[5] She evidently has a general plan, which is to lure the aged caribou into the stream, where she knows she will have the advantage. The grizzly taunts the old caribou repeatedly to induce him to charge her as she sure-footedly backs into the water. Finally, she succeeds, and once they are both in the water, she is in control. The old caribou's footwork on the rocky bottom is unsure, and the game is all but over. Positioning her huge body so that the caribou cannot regain an upright stance is crucial, and the grizzly lunges exactly into the middle of the lethal rack, twisting the caribou to tip him into the deeper water. He drowns, kicking fruitlessly. At no time does the prey make it easy, and the grizzly is constantly ready to modify her strategy.

Or consider wolves cooperating in a pack aiming to bring down an aging elk. Five or six wolves skillfully coordinate their moves, cutting him out from the herd. Those at the elk's rear dodge the kicks and wait for an opportunity to lame the elk by tearing his leg tendons. Those in front harass the elk to distract him from the attack behind. The instant the elk is lamed, those in the front rip his throat out. In these cases, the predators begin with a general plan, while the details of bringing down this particular prey are worked out on the fly. Broader experience enhances how those details are addressed the next time around. Old mistakes are avoided, new opportunities seized. Skills are honed.

In hunting and foraging generally, there is a rich interplay of brain states: motivation, planning, knowledge of the geography, selecting which past experiences are relevant, constant error correction, recognizing the intentions of the others in the hunt, signaling those intentions to others, and probably some moment-by-moment problem solving. The very young watch but do not participate, while adolescent animals stick to the safer jobs until they have learned enough to take riskier roles. Some instinct is involved, but a lot of learning builds on that platform.

What are the neurons doing while we learn from experience? With a host of factors in play, such as memory and motivation and causal models of the world, solving the puzzle of mechanisms for reinforcement learning can seem far out of reach.

In talking strategy in research, Francis Crick was fond of emphasizing that in science you need to attack a problem where you can make progress. Often, at regular afternoon tea in Terry Sejnowski's lab at the Salk Institute, Crick would advise us that in the beginning it is worthwhile avoiding the most complex features of a phenomenon, tantalizing though they may be. His words still echo in my brain: Find a simple entry point. Don't worry if the critics say it is not the whole problem or it is too simple. Never mind that. You may make critical progress even so. If you are lucky, this initial progress will open many doors beyond the first. Then, further complexity can be tackled.

Crick's was a very practical approach, and one I've fondly recalled in mulling over the mechanisms for reinforcement learning.

FINDING MECHANISM

To discover mechanisms that support reinforcement learning, the ideal thing would be to find the neural signature of a simple form of learning, such as associating two events. As Ivan Pavlov (1849–1936) observed, initially his dog salivated only when the food appeared. If, however, a bell was regularly rung before the appearance of the food, the dog began to salivate as soon as the bell rang. The dog's brain had learned that the bell predicted food. This process became known as *Pavlovian conditioning* or sometimes just *stimulus-response conditioning*. The Crick strategy would favor starting by exploring the mechanism of this learned link between the bell and food delivery, and then moving on from there. And that is indeed what happened.

The story begins with Wolfram Schultz, who was recording responses from neurons in the midbrains of monkeys.[6] Each neuron puttered along, firing at a low but significant *base rate* when the monkey was just sitting quietly. Schultz noticed that the neuron's base rate of firing shot up (that is, spiked; Figure 3.2) if the monkey unexpectedly received a reward (a squirt of juice). If the delivery of the reward was regularly preceded by a flash of light, then after a few more trials of "light goes on, then juice comes," the neuron increased its firing rate when the *light appeared*. So far, so good. This is Pavlovian conditioning at the neuron level. The neurons associated the light with the reward.

The neurons that Schultz and colleagues were exploring are in a midbrain nucleus (a clump of neuronal cell bodies) called the *ventral tegmental area* (*VTA*). They are a central part of the evolutionarily ancient reward system (see Figure 3.1).

Here was the puzzling thing: once the neuron *regularly* responded

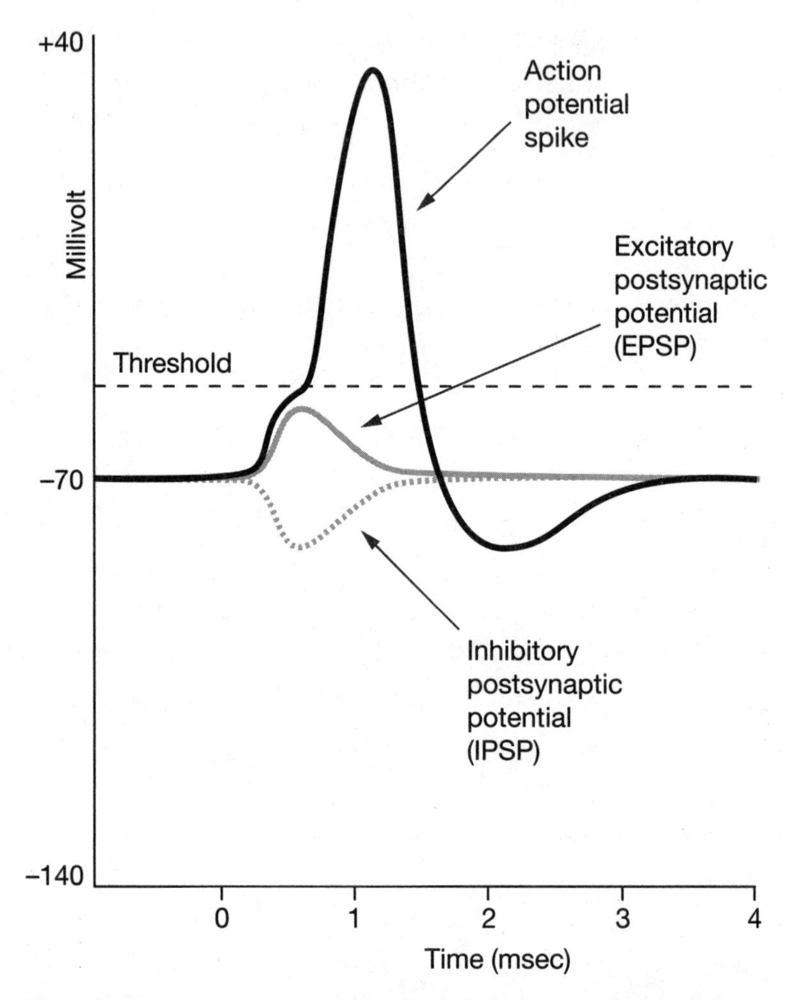

Figure 3.2 What does it mean for a neuron to spike? Each neuron has a voltage difference across its membrane; in this example, the voltage difference is about -70 millivolts. The neuron can receive many input signals (causing small voltage changes) that may converge on the axon hillock at about the same time. Some will be excitatory, and some will be inhibitory. If they add up to a certain level of depolarization of the membrane (the threshold), the neuron will abruptly fire. For the neuron to fire means that there will be a fast, large change in voltage across the membrane of the axon hillock. In this example, the voltage across the membrane reaches +40 millivolts. This change at the axon hillock compels the same large voltage change to travel down the entire length of the axon to its terminal. If you put an electrode into the neuron and record the voltage changes across its membrane, this is what a spike looks like—kind of sharp and pointy, hence *spike*. A spike is also known as an *action potential*. An inhibitory signal hyperpolarizes the neuron, meaning that it has to have a lot more excitation to reach the firing threshold.

to the light, it ceased to respond vigorously to reward delivery, dropping back to its old base rate. In addition, if the light went on but no reward came, the neuron dropped *below* its base rate of firing at the moment the reward was expected (see Figure 3.3). What did these changes in firing rate mean?

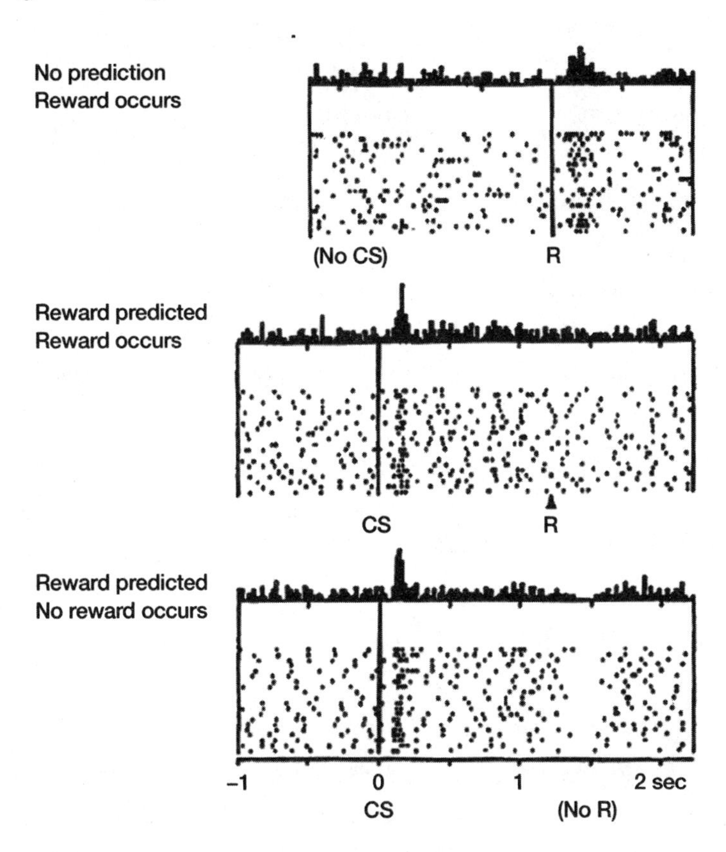

Figure 3.3 Each of the twelve rows shown in the three images is a series of dots, and each dot represents a spike in the neuron. The very top line is a summary (histogram) of the spikes occurring in each of the twelve columns lined up below. Top: The neuron is puttering along at a base rate of spikes, until the monkey gets a juice reward. Immediately there is a brief uptick in the spike rate, then it's back to its old base rate. Middle: After a few trials in which the light first blinks and then after about a second the juice is delivered, the neurons respond with an uptick in spikes only when the light blinks and the reward is expected, not when the juice is delivered. Bottom: If the light blinks but no juice is delivered at the time the reward was expected, the spikes in the neurons fall below base rate. The light in this example is the conditioned stimulus (CS); the juice is the reward (R). The time reference is given in the bottom image in seconds, the whole trace covering about 3 seconds of time. WOLFRAM SCHULTZ, PETER DAYAN, AND P. READ MONTAGUE, "A NEURAL SUBSTRATE OF PREDICTION AND REWARD," SCIENCE 275, NO. 5306 (1997): 1593–99. WITH PERMISSION.

How did this seemingly humdrum result inspire a monumental breakthrough in our understanding of reinforcement learning? To answer, we need first to understand what an unexpected reward means to the brain and why the neurons in the VTA increase their spike rate. At this point the story turns to two postdoctorate fellows, Read Montague and Peter Dayan, in Terry Sejnowski's lab at the Salk Institute from 1991 to 1993. They were consumed by the problem of how reinforcement learning works in the brain. Sharing a computational bent, Montague and Dayan would propose hypotheses to each other, shoot them down, argue about mechanisms, and chew further on the problem. Endlessly.[7]

When the Schultz results were first reported, Montague and Dayan realized that the changes in spike rate of the VTA neurons when the monkey unpredictably gets a shot of juice look very much like an error signal. The uptick in spikes seems to indicate to a different circuit that the usual expectation was wrong—an error. Because getting the juice was a better-than-expected error, these neurons in effect said, "Yay!" and upped their base rate of firing. If the light regularly precedes juice delivery, the firing rate to the light onset goes up. "Yay! Juice is coming shortly."[8] After a few repetitions of "first light, then juice," the delivery of juice becomes normal and hence expected. Consequently, the neurons drop back to their usual base rate of firing, even when the juice arrives. In effect, they are saying, "Same old, same old." Crucially, base-rate firing is *not* no signal at all; it is carrying the information that *nothing unexpected is happening*. So, what seemed puzzling to the Schultz lab is not puzzling if the neurons are responding to expectation; in effect, they are making a prediction about what *will* happen and hence respond according to what *does* happen (Figure 3.3).

When the light came on but no juice was subsequently delivered, the neurons again signaled an error. The neuron's firing briefly dropped below base rate. The outcome was worse than expected. (Boo, no goodie.)

Because they had been scouring the literature, Montague and Dayan knew of a computational model for using error signals in machine learning that had been developed by Richard Sutton and Andy Barto.[9] As they quickly realized, the model fit the Schultz data. The closer they looked, the more beautiful was the fit.

Here is how they pulled together the computational model and neuroscience data. Montague, Dayan, and Sejnowski[10] proposed that what these neurons in the VTA care about is the difference between what was expected at a certain time and what actually occurred at that time. They care about change, which makes good engineering sense, since something should be learned when things change.[11] Hence, changes in firing are learning signals. Once you understand that, you can see the basic mechanism of the spike patterns in Schultz's data.

Unfortunately, however, Montague and Dayan did have a problem. Echoing conventional wisdom, Schultz and his colleagues had concluded in their publication that the neurons from which they recorded in the VTA were *not* involved in representing the expectation of reward. Why not? Because the VTA neurons did not sustain their firing uptick during the interval between when the light appeared and when the juice was delivered.[12] There was just a brief uptick at the time the light came on, then firing dropped back to base rate (see Figure 3.3). Why was that a problem? Schultz and colleagues assumed that only if there was a spike uptick *throughout* the interval between light flash and juice could the VTA neurons know *when* the juice was expected. Therefore, they reasoned, absent the sustained spiking uptick, the neurons could not be signaling either expectation of reward or error in expectation. They must be doing something else, such as signaling that attention should be paid.

The conventional wisdom warranting this conclusion had a chilling effect on progress. Montague and Dayan knew that after learning the drop down to the old base rate during the delay was exactly what the Sutton-and-Barto model actually *required*, because nothing unexpected happens during that interval. Consequently, in their submitted paper they meticulously explained how the VTA neuron knew when the reward should be delivered.

How does the neuron get the timing right? Rather simply, as Montague and Dayan figured out. With the repeated flash of a light followed by the reward, the increase in firing of the neuron is transferred to the *earliest reliable stimulus* for the reward. Thus the light becomes a predictor of juice delivery, and thus the time of the light onset and the time of the predicted reward delivery are crucial parts of the mechanism. That seems a lot for a single neuron to learn, but not if it is located in just the right place in a vastly larger and massively complicated bit of neural network machinery. Which it is, as we will shortly see.

It took Montague and Dayan four tenacious years and ten painstaking rewrites to get their interpretation of the results accepted for publication. Sejnowski, their lab chief, knew that their interpretation of the Schultz data had to be right, and he sensibly waved off the repeated rejections as the price you pay for a really good idea. One can safely guess that when the acceptance letter finally arrived, "better than expected" was the response in the researchers' own VTA neurons. Surprising new ideas often endure these setbacks, and in the end, the tenacity of this research team paid off. What had been conventional wisdom found itself listed as yet another teachable example of people-used-to-think narrowness.

That the VTA neurons are signaling reward prediction error matters in the brain only if other neurons get that signal and do something with it. Who is listening? The VTA neurons broadly send their axons to another region also in the ancient part of the reward system: the basal ganglia, or more exactly, the *nucleus accumbens* in the basal ganglia.[13]

When a spike on a VTA neuron reaches its target in the nucleus accumbens, the axon terminal releases a neuromodulator, dopamine (Figure 3.4), which acts as a learning signal: "Do that again." If the

VTA neurons are spiking *above* base rate, they will release more dopamine than when they are *at* base rate. When the VTA neurons do not spike (that is, when the outcome is worse than expected), they release nothing.

Next, the released dopamine binds to specific receptors in the accumbens neurons. This action changes how the accumbens neurons behave. Some neurons in the accumbens are involved in action choice. Others, intriguingly, are associated with feelings of pleasure. Some of the accumbens neurons have receptors for opioids or for cannabinoids. (As noted earlier, cannabinoids are marijuana-like neurochemicals made by the brain. Opioids are opium-like neurochemicals made by the brain.) When cannabi-

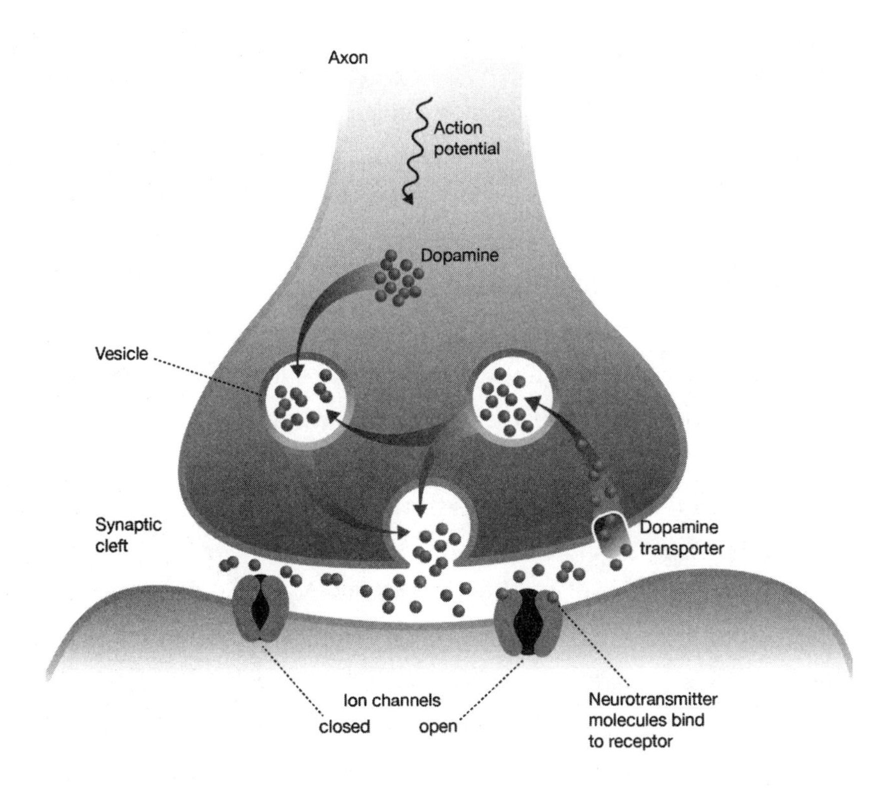

Figure 3.4 Simplified, the organization of the synapse, showing the mechanism by which neurotransmitters (in this case dopamine) are released.

noids or opioids dock on their specialized receptors, they make us feel good; that is, they generate positive feelings. So, here in the accumbens we find a link between learned value (positive value) and the VTA pattern of dopamine release.[14]

An additional subtlety emerges from research on the dopamine receptors themselves. The nucleus accumbens has at least two types: D1 and D2. *Enhancement* of positive reward learning is supported by neurons fitted with D1 receptors, whereas *diminution* of positive reward learning (sort of "don't get too excited yet") is supported by neurons fitted with D2 receptors. Among the general population of dopamine receptors, therefore, is a kind of balancing between the two competing subpopulations, perhaps making abrupt changes in value prediction less likely, unless the positive reward is huge.[15]

In short, VTA fluctuations in spike rate, as when the animal gets a nice reward, cause fluctuations in dopamine in the accumbens. These fluctuations influence action choice: go for it, or don't bother. In mammals, the basal ganglia, including their nucleus accumbens, are intricately connected to specific regions of the frontal cortex. The VTA neurons themselves also directly send broadly distributed dopamine signals to neurons in the frontal cortex. One region, the orbitofrontal cortex, seems to update the valuation of events in response to dopamine signals.[16] There are also loops from the frontal cortex back to the basal ganglia and the midbrain. Knowing who talks to whom does not mean we know precisely what is being said. We do not. But identifying the pathways is a crucial preliminary to figuring out the neural mechanisms.

So far, this account is about positive reward and its absence. What about a painful outcome—call it *negative reward*, to maintain the parallel? Suppose you steer your speeding bicycle to a gravel side road, and you skid out. You end up with bleeding knees and arms. Much worse than expected. Next time, you avoid gravel or greatly reduce your speed or both. Are penalties signaled in a manner opposite to reward? Pretty much.

Penalty-sensitive neurons increase activity when the outcome is much worse than expected, and they decrease activity when the outcome is better than expected. My brain's lesson: don't steer a bicycle into gravel. As Okihide Hikosaka and colleagues at the National Eye Institute discovered, these penalty-sensitive neurons originate in a small structure called the *habenula.* A pathway from the habenula reaches the VTA in the brainstem, where its main function is to inhibit the VTA neurons and suppress a motor response that the habenula neurons deem harmful. Their message is "Probably this should be avoided." Whereas the VTA neurons release dopamine, the neurons in the habenula release serotonin.[17]

A different way of describing the meaning of the VTA neuronal activity is this: it reports the value of a future event—for example, whether the value of what can be expected to happen soon after the light goes on is something worth reaching for, worth taking a risk for, or worth pursuing. In this manner, learning and deciding are linked. The wider and deeper the sensory experience of the world that is available to the basal ganglia, the more complex is the evaluation of what to expect and how to optimize the positive things. In a social context, the brain learns social values. We get disapproval for fibbing, and approval for waiting our turn. Our brains get a big reward boost (dopamine hit) from approval. Our brains get a big serotonin hit for disapproval. To a first approximation, this is the means whereby our conscience is configured.

Reward prediction error seemed to be one of those entry points that Crick envisioned. Coding for reward prediction error has the earmarks of the critical component in all reinforcement learning, not just Pavlovian conditioning. Unlocking more doors is possible in mammalian learning if it turns out that the basic mechanism is integrated with neocortical embellishments and extensions that could explain the kinds of reward learning that are more sophisticated than vanilla Pavlovian conditioning—that is, if neurons in the VTA and basal ganglia are smoothly knitted into the evolved networks in cortex and in the hippocampus. Lo and behold, they are. The loops between the basal ganglia and the cortex are rich and ramified in all mammals.

Although the simple case is not all that simple once you delve into the cellular details, it does provide the means to address more complex forms of reinforcement learning. Beyond Pavlovian conditioning there is *instrumental conditioning*, or learning by doing, not just by waiting, watching, and associating stimuli.

Dogs learn to pull a bell rope when they want to be let out of the house; rats learn to press a lever for food pellets; human toddlers learn to open the gate by lifting the latch and pushing, and they learn to get a toy from a shelf by pointing at it while vocalizing. First we explore our situation, then we either repeat the action sequence if the attempt worked out well or modify it if it did not. We learn from our mistakes and from our victories, and even from our no-change-needed results. Explore and exploit, as the lore has it. These are all instances of reinforcement learning that involve the usual suspects: dopamine, serotonin, the nucleus accumbens, the VTA, and the prefrontal cortex.

One "techy" point to round out this section: The computational model that Montague and Dayan used to explain basic reinforcement learning in brains is essentially the very computational format used in machine learning, or deep AI (artificial intelligence), now the darling of the techy world.

In machine learning, a computer network can learn to perform complex pattern recognition tasks such as face recognition. Unlike conventional computing, however, the network is not coded for its pattern recognition job. It has no program. It is an artificial neural network, with pretend neurons and pretend synapses connecting them. It learns from exposure to examples. How? By reward prediction error mechanisms. Just like the VTA and the nucleus accumbens. After exposure to an example, the machine suggests an answer, to which it gets feedback—yay or boo. Depending on the answer, small changes are automatically made to the pretend synapses between the pretend neurons in the network, a bit like small changes made in the nucleus accumbens and cortex in response to dopamine fluctuations. The artificial neural network learns by trial and error.[18]

Reward prediction error is the protocol in AlphaGo, the artificial neural network (ANN) machine-learning device that learned to play Go and beat the top Korean player, Lee Sedol. It also governs learning in the ANN that plays Texas Hold'Em, demolishing world-class poker players.[19] And it is the same protocol that is used in teaching ANNs to recognize suspicious cells in a mammogram. The engineering strategy of mimicking reward-prediction-error mechanisms in the brain turns out to be vastly more flexible and powerful than the conventional code-a-program strategy.

BEYOND SIMPLE CONDITIONING

Learning by doing can become remarkably fancy, especially when it draws on intricate background models of how some part of the world works. In growing crops, such as raspberries, background knowledge of gardening, however acquired, comes into play. You learn how to recognize overwatering and underwatering, and when fertilizer would give the plants greater productivity. Pruning is an art, with little in the way of highly specific rules; it's roughly "not too much, not too little." So, you experiment until you get the hang of it. In raspberries, fruit is produced on the second year's growth. When pruning in the fall, the question is how much to cut back the current year's growth to optimize the plant's productivity the next summer. This means you do not know the results of a pruning strategy until the following year. To confound things a bit, other factors may affect productivity, such as lack of fertilizer, or pests. Simple instrumental conditioning is obviously inadequate for this learning task. Background knowledge and a good memory are essential in expanding the power of reward prediction error.

Self-control, dependent on regions of the frontal cortex, is crucial in inhibiting suboptimal choices, such as choosing immediate gratification, with the result that you forgo better, long-term rewards. Roughly, the more neurons in the frontal regions, the greater the capacity to control impulses. Even rodents, however, with their rather modest prefrontal cortex, can show impressive self-control.

Here is how we know that. Rats easily learn that pressing Lever A yields one food pellet, but pressing Lever B produces *five* pellets. We give the rat only one press per trial, so it has to make a choice: either A or B. The obvious better choice is B. Now suppose the experimenter introduces in B a delay between lever press and pellet delivery. Some rats will choose B even if the time between lever press and pellet delivery is extended to 30 seconds. They wait and optimize their reward.[20] As in humans, however, there are differences between individuals in their capacity for self-control.[21] Some rats, like some humans, have poor control over their impulse for immediate gratification, usually going for the lesser but more immediate reward.

In humans, it has been shown that acute stress modifies the connectivity between cortex and basal ganglia, with the result that the value of the immediate reward (e.g., cookies) is enhanced, while the value of the longer-term preference (e.g., cheese) fades.[22] Other factors, such as exhaustion and fear, can also affect self-control.

One complication in the simple reward-prediction-error story involves changes *internal* to the animal's nervous system that alter the profile of what is rewarding and to what degree. Juice will not be as rewarding when I am sated with juice as it is when I am thirsty; highly salted liquids will seem bitter and to be shunned, unless my salt levels are abnormally low, when I will seek salty liquids and enjoy the taste.[23] If I am exceptionally stressed, I may not find food very rewarding. Many heroin addicts report that they no longer find heroin rewarding, but they take it to ward off the miserable effects of withdrawal. They may still have a powerful desire for the drug, but they no longer say they like it. Some say they hate it, for what it has done to their brains. Wanting and liking can come apart. Some part of the reward system and its connections undergoes changes in addicts.[24]

SIMULATING OPTIONS

Should I go fishing or play golf? Should our children change schools, or piano teachers, or soccer teams? Humans, and probably some other animals, often make plans by simulating and evaluating the outcomes of possible actions, the choice being influenced by the better outcome for that individual in those specific conditions. This process is called *prospective optimization*.[25]

The same mechanisms involved in immediate evaluation (choosing either the carrots or the candy bar) are also involved in more deliberate evaluation of simulated outcomes (going to college at UC Berkeley versus UC San Diego). In the mix of brain processes that we call deliberation are the recollection of similar cases, visual imagination, relevant factual knowledge, self-knowledge about one's own preferences and character, and who knows what else exactly. Without doubt, the capacity for simulation and evaluation requires cortical machinery and subcortical machinery, but details of how brains simulate nonactual events are still largely unknown. In a general sense, however, optimizing by evaluating relevant options, and deploying impulse control in selecting what is deemed overall best in the medium to long run, seems to be the procedure. This process is also called *constraint satisfaction*, as we will see in Chapter 7.

WHAT I SHOULD HAVE DONE

Counterfactual error involves a different kind of learning from experience. An example is buyer's remorse, where we recognize that, of the options available, the choice we made was worse than the option or options declined. I still recall my first car, the Austin Devon, as one such unfortunate choice. This counterfactual judgment requires tracking the outcomes of both the choice made and the choice(s) forgone, and comparing their values. The Austin's clutch failed in the first month I owned it, whereas the Nash Metropolitan that cost a hundred dollars more was still repair-free three years later.

Terry Lohrenz, a neuroscientist collaborating with Read Montague, realized that counterfactual learning in humans can be explored in an experiment in which subjects are given real money to make investments in the stock market game. (He used an old, but actual, stock market chart.) At some point, each subject can make a monetary investment—either cash or stock.[26] Once the subject places the bet, the actual direction (up or down) of the market is revealed. If the market goes up and you chose cash, the value of your portfolio is lower than it would have been had you chosen stock. You experience a bit of regret. As well, you learn that investing in cash is not always optimal. When you again have the opportunity to choose between cash and stock, your past regret is apt to make you a little more likely to choose stock.

When subjects made investment choices while their brains were scanned for changes in activity levels, the data did indeed reveal sensitivity to counterfactual error in the nucleus accumbens.[27] The results indicated that both experiential and counterfactual error are involved in real-world decisions. We frequently reassess whether our actual decision (cash) was optimal or whether we should have taken the other option (stock). Keeping track and comparing both kinds of errors involves smart, cognitive operations. As Lohrenz points out, evaluating counterfactual error involves rewards *not* received from actions *not* taken. This means that a subject's evaluation of counterfactual error has to draw on background knowledge and the capacity to organize that knowledge so as to make a reasonable assessment of the cost of the action not taken.

From the reward system research in monkeys and rodents, we might infer that this signature seen in the fMRI image reflects VTA dopamine release on accumbens neurons—boosting or depressing activity. To test that inference in animal models, probes are painlessly inserted into the animal's basal ganglia. In humans, however, invasive probes cannot be used without a proper clinical justification, so the inference, even though plausible, needs to be verified with human data. This is not just to satisfy our curiosity, but because the various addictions, as well as psychiatric conditions such as depression and schizophrenia, are associated with dopamine dysregulation. Consequently, it is highly desirable to know as much as possible about the reaction of neurons in the reward system in the human brain.

In a landmark study published in 2016, Kenneth Kishida and colleagues found a way to obtain relevant data that was both ethically acceptable and scientifically clever. They discovered that in humans, the activity in the nucleus accumbens is indeed coupled to dopamine release—up for positive reward, down for negative reward. To their surprise, however, the encoding turned out to be more subtle than pure and simple reward prediction error.

Their experimental subjects were human patients who were to receive deep brain stimulation (DBS) treatment for advanced Parkinson's disease. DBS is now a common and generally effective intervention for severe cases of Parkinson's. To reach the "deep" site (the subthalamic nucleus), an electrode approximately the diameter of a small knitting needle is safely inserted down through the cortex. A surgical team at Wake Forest University Health Sciences implanted the electrodes in the patients. As it happens, the preferred insertion route passes very close to the nucleus accumbens. When he realized this, Kishida decided to engineer superfine diagnostic electrodes to accompany the treatment electrodes. His colleagues at Wake Forest University Health Sciences agreed to collaborate, and seventeen patients consented. Once in place, these electrodes could collect data on subsecond dopamine release in the accumbens.

When the patients recovered from surgery, they were ready to play the stock market investment game. The measurements of dopamine release began. As the patients made investment choices, their dopamine fluctuations in response to better- or worse-than-expected outcomes were recorded. The first of its kind, this experiment sets a benchmark for future work.

The results had an interesting subtlety. With large bets, the dopamine fluctuations were as predicted: more dopamine when the outcome was better than expected, less when worse than expected. When the bets were small, however, the reverse effect reliably happened: more dopamine to a small *loss*, and less dopamine to a small *win*. This pattern had certainly not been predicted. What were the human brains encoding when the wins or losses were small rather than large?

The hypothesis to explain this effect says that measuring dopamine levels in the nucleus accumbens of humans is equivalent to measuring an integrated result of both reward prediction error and counterfactual error. In short, when a subject has experienced a small loss, his reward system also responds to a counterfactual condition—how much better his win could have been had he chosen otherwise. As the patient contemplates this counterfactual big win, his reward system is sort of responding, "Ah yes, it could have been so lovely." It is a bit like thinking how good you would have felt if your basketball team had scored the buzzer-beater shot and won. Merely the contemplation of a win feels rather good, which, I suppose, is linked to why we often indulge in rewarding fantasies. I still think about that Nash Metropolitan I did not buy.

The Kishida explanation, if correct, suggests that although some

VTA neurons are hewing to the basic reward prediction error template, some are computing counterfactual values, and some accumbens neurons are causing a hedonic (positive) response to an anhedonic (negative) experience. A finer level of neuron-by-neuron exploration of VTA neurons in humans, were it feasible, should reveal this division of labor. Perhaps this division of labor also exists in nonhuman mammals but was not picked up in earlier recording of dopamine levels in the nucleus accumbens. This is another reason why finding a way to obtain data on the human brain is essential.

Dopamine is not the only player here. As we saw earlier, serotonin is released when a choice turns out badly. It is the brain's way of telling us to watch out for that choice henceforth. So here is a question: In the investment game, when the subject makes a big bet on stock but the market goes way down, dopamine will not be released (because the outcome is worse than expected), but what does serotonin do?

Ken Kishida and Rosalyn Moran asked this question and used Kishida's diagnostic electrodes to track serotonin release in the nucleus accumbens while Parkinson's patients played the stock market game. Thus they could compare the distinct functions of dopamine and serotonin in the same region and during the same task.

Serotonin release is the opposite of dopamine release—it goes up for bad decisions, and down for good ones—but it is also sensitive to whether the *actual* loss is large or small and to whether the *counterfactual* loss is large or small. Things get really interesting when events cause both dopamine and serotonin release. For example, dopamine can go up in response to a big counterfactual win, while serotonin simultaneously goes up owing to the actual loss. In such instances, subjects respond like this: "Hmm, well, next time I will not put so much in the market—just a small amount."

As Moran put it, "Serotonin acts in a way that reminds us to pay attention and learn from bad outcomes, and to promote behaviors that are less risk seeking but also less risk averse. When there's an imbalance of serotonin, you might hide in a corner or run towards the fire, when you should really be doing something in between."[28] Serotonin, Moran and Kishida suggest, is the "keep calm and carry on" system, preventing us from overreacting to both negative and positive outcomes.[29] The balance between the two modulatory systems—dopamine and serotonin—is exquisite and may well have a lot to do with the kinds of balance we try to achieve in life.

Are rodent brains capable of counterfactual evaluation? Probably. Some experiments with rodents indicate that in the deliberation period prior to making a choice between two alternatives, neurons in the prefrontal cortex (orbitofrontal, more exactly) alternate activity as the rodent turns to study one choice, then the other. After a selection is finally made, the rodent brain, too, can signal regret at not having chosen more fortunately.[30]

Humans may have the capacity to look further into the past than other primates can in assessing counterfactual error. We ruminate about what might have been, had we played the flute instead of the trombone in middle school, or had we bought a Nash Metropolitan as a first car instead of that Austin Devon wreck. We also ruminate about counterfactuals regarding others: How would things have turned out if my father had been able to go to college instead of apprenticing as a printer's devil? Some counterfactuals take us so far from the world of facts that they are just pure fictions, and hard to ground one way or the other: Would communism have been successful in Russia if Stalin had been a morally decent human?

Another variation on the reinforcement learning theme is known as *learned industriousness*. It is well known that the pleasure from the rewarding event may transfer to an action that typically produces it. If great physical effort is required to achieve a reward, such as in splitting firewood, the sensation of making the strenuous effort acquires a secondary reward that somewhat mutes its normal aversive effect.[31] Splitting wood is a tough job, but some old hands split wood just for the sheer joy of it, even if the woodshed is already amply stocked with firewood. Often, habits initially acquired when the reward was regularly produced (a ready pile of firewood) continue long past the fading of the reward. Something like this lies behind the habits of workaholics.

Reward learning is also part of learning how to learn. Suppose you find that solving mathematics problems or practicing the piano is onerous and hence you tend to procrastinate. One strategy is to make a deal with yourself whereby you work steadily for half an hour and then give yourself a nice reward, such as twenty minutes of video gaming. After a few such trials, the resistance to getting down to work fades, and you may begin to find some enjoyment in the work itself. This way of exploiting the reward system is an instance of the old-fashioned but dependable mantra "chores first, play later," which many parents rehearse until it is well ingrained.[32]

COGNITIVE PATTERN GENERATOR

We learn, partly through experience and partly through instruction, how to organize a sequence of actions to extract a splinter or change a tire. Although proper sequencing is a difficult computational problem for brains, they have evolved to handle such problems remarkably efficiently. Ann Graybiel, a neuroscientist at MIT, discovered that the basal ganglia contain clusters of neurons whose activity is so orchestrated that the right sequence of activity is produced when we perform a multistep, semihabitual action.

Graybiel realized that alongside motor skills such as pitching a tent, and habits such as taking the same route home every day, there is a wide range of complex cognitive functions that are sort of skill-like and sort of habit-like, but that draw on extensive background knowledge, both of a general sort and of a specific sort. She perceived that for regularly encountered motor goals, such as riding a bicycle, motor pattern generators output motor sequences. Her additional insight was that, in a similar fashion, there are cognitive sequences that our brains deploy as we re-encounter cognitive problems previously solved successfully. For example, an experienced nurse can efficiently and appropriately triage patients in the emergency room.

To handle these sorts of cases, Graybiel devised a breakthrough concept: the *cognitive pattern generator.*[33] How do you teach basic logic to freshmen? My cognitive pattern generator worked out a smooth pattern over many years. I have to be awake and alert to teach the class, of course, but the sequence of steps flows almost effortlessly. Having found specific examples that clearly and succinctly get the point across, I notice that those very same examples will come out of my mouth, year after year. Consider another kind of problem: how to argue a libel case in court. A veteran trial lawyer's cognitive pattern generator kicks in, and once again she smoothly goes through the steps to square away the details and do what needs to be done. This capacity for cognitive pattern generation is not unique to humans. To recall an earlier example, if you are the grizzly aiming to bring down the old caribou, how do you go about it? The experienced grizzly well knows the general pattern for successfully bringing down the caribou, and past success gives her confidence in her knowledge.

Are there social problems whose resolution benefits from a cognitive pattern generator? Such social problems might include how to deal with a fractious colleague, how to galvanize a lazy but gifted graduate student, or how to prepare an anxious patient for a risky surgery. In these kinds of cases, we deploy cognitive patterns shaped by the reward system over years of experience. Just as with motor pattern generators, we customize the sequence to fit the case at hand, but the general form of the learned cognitive sequence may be exactly what is called for. (See also Chapter 7.)

Cognitive patterns are efficient because once you size up a situation, you can transfer a well-honed cognitive pattern to a new instance. Just as you can transfer your bike-riding skills to mountain bikes and fat bikes, so can you transfer your cognitive skills, with suitable adjustment, from one case to another rather similar case. Your cognitive pattern generator can take you from dealing with a fractious colleague to dealing with an obnoxious boss; it can take you from coping with a petulant aunt to coping with a cantankerous father-in-law. Sometimes the cognitive pattern involves rather simple sequences, such as seeking advice from a trusted person, or maybe resolutely doing absolutely nothing.

Cognitive pattern generation is a breakthrough concept because it gives us the tools to see how cognitive skills of many different kinds, including those that are needed for social problem solving, are acquired by a highly cooperative basal ganglia and frontal cortex organization. Recognizing the role of the reward system in cognition also gives us clues about why rituals reduce anxiety and how rituals can sometimes become problematic habits.[34]

Internalizing norms involves the VTA and the nucleus accumbens, but what happens when the norms we live by shift somewhat? Does the reward system signal *norm prediction error* and then we change accordingly? The next chapter will look at the evidence suggesting that our brains use norm prediction error to change their norms.

CHAPTER 4

Norms and Values

In fact, to have no sense of humor is to be a seriously flawed human being. It's not a minor shortcoming; it shuts you off from humanity.
ALAN BENNETT[1]

SOCIAL LEARNING AND SOCIAL BONDING

We were rafting the Firth River, far north in the Yukon Territory of Canada. On this two-week expedition were eight undergraduate students from UC San Diego, two professional guides, and me. For the students, rafting deep in the Canadian wilderness was totally new. All were greenhorns. As we approached Class 4 rapids in a canyon, it was decided that, rather than portaging the three heavy rafts and all the gear up the canyon sides and down again, we would line the loaded rafts through the rapids and then hike downriver where the water was manageable. Lining a raft is a bit like walking a dog; the more wild the rapids, the more unruly the task.

Lining entailed close cooperation among everyone. We were stationed atop the canyon rim, holding ropes tied to the bow and stern of each raft resting in the river. The task required one group to walk carefully some distance downriver and then hand off the ropes to the next student in the line and go back up the line to take over the ropes for the next raft. The procedure took about three hours and was not without danger, given the steep canyon walls and the turbulent river. In the event of an accident, we might, with luck, make radio contact with the outside world, but we were essentially on our own. If a raft were lost, food and gear would be lost. Everyone had to perform their job exactly so, or they stood to imperil the others.

At the end of the trip, each student wrote about the experience they most valued. Although they had sneaked belly-down to observe a herd of musk oxen and explored a glacier, it was the day of lining the rafts that stood out for them. Without much thought, I had assumed they would find it mostly a rather strenuous chore. I was wrong. They found the day of intense cooperation and danger to be especially memorable and satisfying. They were moved by the simple and compelling nature of the cooperation and their mutual dependence. The levels of laughter and fun went up after the lining of the rafts. They all felt much closer to each other.

To redescribe this incident from the perspective of the brain, cooperation with their peers in the teeth of nontrivial danger was strongly rewarding. The VTA neurons in the students' brains were almost certainly sending gobs of dopamine to their targets in the nucleus accumbens. Endogenous cannabinoids were released hither and yon. Among many other neural events, of course. Significantly, the interactions of the students on the rest of the journey revealed ever-stronger friendships and joy in each other's company.

Play has a comparable effect. Social bonds strengthen, and learning to win and lose with grace provides a template for how to behave in more serious social situations. Cooperation in play teaches the basic lesson that "in union there is strength," and that many accomplishments are possible only if we work together. In play, unfairness is recognized, and patterns for dealing with it emerge. Bossiness gets discouraged, understated leadership is encouraged. Certain responses get approval, others get disapproval. All the while, the reinforcement learning system, both in its subcortical and in its neocortical aspects, is changing the brain—its preferences, biases, patterns, and expectations.

Social learning in humans involves a lot of watching and imitating and trying on your own. At a wedding, the little ones rarely need much encouragement to join in the dancing. They watch for a minute or two, and then get up and dance—tentatively at first, then with more confidence. Typically, children intently watch and pick up styles of behavior such as kindly interactions, generosity, warmth, and friendliness, as well as their opposites. Children model their reactions on what they observe in family and friends. When a social situation is deemed especially unsavory, explaining why we're not helping a mischief maker will add to a child's social sophistication. Frequently, a child gets such explanations by listening in as the adults discretely discuss the matter. Listening in provides information that a child often cannot get directly—a reality that adults often understand but may not acknowledge.

Imitation and cooperation both feel good. When the neighbors gather to milk the cows of a dairy farmer stricken with the flu, the children watch and pitch in. They pick up on the fellow feeling that comes from cooperation. Without being asked, they put hay in the racks, fill the water troughs, and crank the milk separator.[2] The reward system is positively engaged.[3]

Humans are clearly prodigious imitators, though how broadly nonhuman animals and birds can imitate remains somewhat controversial. It is well known that in many bird species, young male birds learn the song of their father by imitation, and the neural mechanisms have been well explored. Capuchin monkeys are intensely social, and highly imitative. They rely on transient coalitions for a variety of tasks, including foraging, and they have communicative rituals that are learned.[4] Parrots, famously, have stunningly powerful imitative capacities, mimicking many speech sounds and many animal sounds, such as those of wolves, cats, and pigs, as well as whistling for the dog, burping, coughing, and sneezing.[5] Jane Goodall and her colleagues videotaped young chimps learning from their mother how to break open a large fruit. They watch, try it for themselves, and then learn from their errors to improve their technique.

Along with carefully documented results, lots of anecdotes attest to the imitative capacity of nonhuman mammals, and many involve dogs. My dog Farley is an eye-contact dog. That means he is fond of gazing into my eyes while I gaze back, nuzzling his ears and talking to him. It turns out that when I talk this way to Farley, I also smile a lot. One day, midsentence, I realized that Farley was smiling back—upper gums pulled up, teeth in view. It may not have been beautiful, but it was endearing. In humans, social smiling begins in infancy, and it appears to be highly rewarding to both adult and infant.[6]

Are there perhaps two reward systems, one for social reward and one for nonsocial reward, each dedicated to its special domain? Probably not.[7] In their thorough analysis of a vast array of existing data, covering many techniques, neuroeconomists Christian Ruff and Ernst Fehr show that the evidence strongly points to just one system, with one set of interlocking mechanisms for assigning value—positive or negative.[8] Depending on whether the context is social or nonsocial, differences do emerge in relevant wider circuitry. Social representations, for example, will tap into knowledge bases involving recollection of social customs, as well as specifics about individual characteristics, but such knowledge will play little, if any, role in strictly nonsocial decisions (Figure 4.1).

The idea that the basic mechanisms for assigning value are common but the knowledge pools accessed for specific actions differ makes good evolutionary sense. Repurposing and refining an existing operation is a typical evolutionary maneuver, and whereas designing a complicated mechanism from scratch might work for an engineer, it is not how biological evolution works.

To take a minor example of the common-mechanism hypothesis, if partway through dinner in Japan, I am quietly instructed that

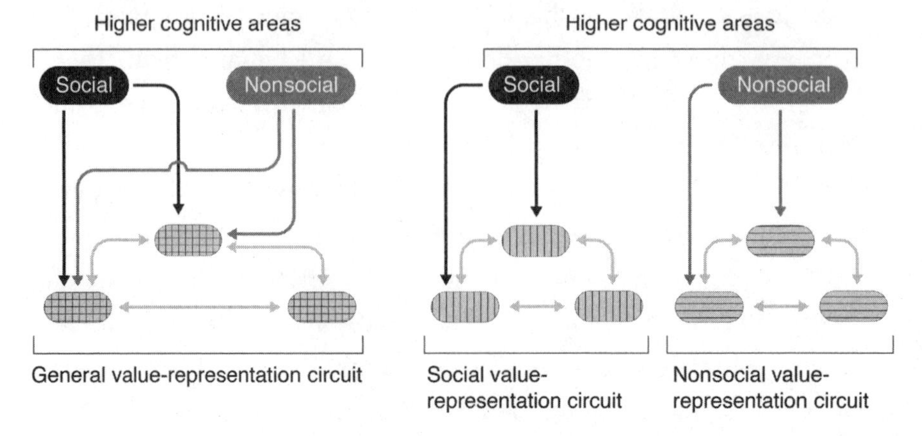

Figure 4.1 Two competing hypotheses for how the brain may determine the value of either social or nonsocial factors during decision making. Left: The common-currency schema assumes that a single neural circuit (shown as hatched) determines the motivational significance of both social and nonsocial events. The perceptual and cognitive information that is relevant to these unified value computations may differ between social and nonsocial choices and may be provided by distinct brain areas (shown in black and gray). Right: The social-valuation-specific schema proposes that social aspects of the environment are processed in a neural circuitry that evolved specifically to deal with social demands (vertical lines). The evidence supports the common-currency model. ADAPTED FROM CHRISTIAN C. RUFF AND ERNST FEHR, "THE NEUROBIOLOGY OF REWARDS AND VALUES IN SOCIAL DECISION MAKING," *NATURE REVIEWS NEUROSCIENCE* 15 (2014): 549–62. WITH PERMISSION.

noisily slurping noodles is expected as a sign of enjoyment, my social knowledge is enhanced. By contrast, if I learn from experience that dry wood is harder to split than green wood, no knowledge about the social world is involved, and that fact is stored in a different region of the brain. The main point is that the paths to knowledge domains in each case are different, and likely the accompanying feelings are also different. One involves social embarrassment, where I wonder what you think of me and whether my reputation has taken a small hit. The other, not. But the core mechanisms for assigning value are likely the same.

On occasion, solving a nonsocial problem may have a virtual social dimension. What will my neighbor think of my raspberry patch—poorly pruned, or well maintained? Despite myself, I tend to care, at least a little. We humans are intensely social, even when we are doing or learning something not explicitly social. Our reputation matters to us for powerful social reasons, and how we deal with non-social facts sometimes impinges on our social reputation.[9] For example, we are apt to regard someone who insists that the Earth is flat as a bit unreliable and hence as unsuitable for school superintendent.

Vicarious social pain and pleasure also activate our basal ganglia and frontal cortex. Suppose that you must undertake the unpleasant task of having a chat with an underperforming graduate student who needs to think about a different career. When I learn of your unpleasant chore, I feel social pain, albeit more mild than yours.

Empathy is an ever-present factor in how humans decide what accords with the demands of conscience. In humans and perhaps all highly social mammals, an empathic response comprises a cluster of functions, including the capacity to cognitively appreciate the situation of another, the capacity to take the other's perspective on a situation, and the capacity to match, perhaps in a diminished form, the emotions experienced by someone else. (In Chapter 2 we saw that if one prairie vole is stressed, the partner's stress levels quickly rise to match those of the stressed vole, and the partner begins intense licking and grooming, which raises oxytocin levels, thereby reducing anxiety.)

Empathizing is not a single operation, in contrast to, say, an eye-blink response to a puff of air. The multiplicity of functions means that each of them can be shaped and influenced with some independence from the others. And indeed, that does happen. For example, some people are adept at cognitively determining what others feel but rarely match those feelings themselves, perhaps because they have been trained that way. For example, doctors and nurses responding to agony in the emergency ward may need to scale back their tendency to approximately match the feelings they see afflicting their patients. Otherwise, their emotions may overwhelm their capacity to cognitively size up what is needed to care for the patient.

A very sympathetic person may tend to swiftly match feelings with the aggrieved, without taking time to consider more dispassionately what may really be going on and whether the empathy is deserved or possibly owed to manipulation. Getting the whole story behind why a colleague feels wounded is often essential if we are to avoid being manipulated.

Empathy varies across individuals and within an individual, and across types of external conditions as well as internal conditions. Yes, when conditions permit, humans frequently do empathize with the plight of others. Nevertheless, as the psychologist Roy Baumeister realistically reminds us, "The key point is that this empathic sensitivity seems to be selective. People may feel a great deal in some situations and towards some targets, but they seem to lack it utterly in others. And people are surprisingly flexible in their capacity to feel sensitive and empathic towards some and not others."[10] Moreover, most of us do not have an infinitely large capacity for empathy. We falter under empathy fatigue.

There is evidence that, on average, females are more empathic than males, but I hasten to emphasize *on average*, and in any case the effect is larger for emotional empathy than for cognitive empathy or for empathic responses that are expressions of social skills.[11] Until recently, little research focused on the question of heritability of each of the various empathic factors in males and females. A recent twins study involving seventeen hundred twins in Italy indicated that there was little difference between identical (monozygotic) and fraternal (dizygotic) twins in heritability of the various psychological factors that make up the whole empathy bundle, though the analysis suggested that genetic factors might be more important in females than in males for emotional empathy.[12] This result is preliminary, and no firm conclusions can yet be drawn about heritability.

Many areas of the cortex, especially those in the frontal regions, appear to be associated with empathic responses, but until we have much more data that are relevant from the cellular and network levels, the nature of the mechanisms will remain unclear.

SOCIAL NORMS AND EXPECTATIONS

Social norms—governing such matters as fairness and lying—guide our expectations and our reactions. When students discover that one of their number cheated by stealing the exam answers, they raise the cry of unfairness and may ostracize the offender. When the professional cyclist Lance Armstrong finally admitted to doping for years, the sports world roundly condemned the unfairness of his seven Tour de France wins, and he is now widely snubbed, as well as banished from cycling competitions.

An unfair allocation of food is clearly appraised by monkeys, as Frans de Waal and his colleagues have carefully shown.[13] The monkeys in the experiment like both grapes and cucumbers, but they prefer grapes. The monkeys have a basket of small pebbles in their cage, and to get a treat they must give the experimenter a pebble in exchange. In the test for sensitivity to unfairness, one monkey exchanged its token and got a grape, while the other monkey exchanged its token and got a slice of cucumber. They both saw what the other received. Outraged, the cucumber monkey accurately pitched its cucumber slice directly at the experimenter. De Waal and colleagues have also shown that chimpanzees who have to cooperate to obtain food will punish a freeloader,[14] not unlike humans who register their negative attitudes regarding freeloaders with a kitchen sign saying something like, "Them that works, eats."

A rather perplexing result regarding violation of fairness norms turns up when people play a simple fairness game called Ultimatum. In the game, one human subject, call him Donor, is given a sum of money—let's say ten dollars—by the experimenter. Donor can make an offer between zero and ten dollars to another subject—call him Responder—who can accept or reject the offer. If Responder accepts Donor's offer, then Responder gets what was offered, and Donor keeps the rest. All done. If Responder rejects Donor's offer, however, neither gets anything. Zero dollars for both. In the canonical case, neither subject knows the other or even sees the other, and they play the game exactly once. An important variation consists in multiple rounds of donating and responding in the game.

Economists who think about rationality are quick to emphasize that it is always rational for Responder to accept an offer as low as even one dollar, and always irrational to reject any offer, however small. It is always rational for Responder to accept one dollar because one dollar is always better than zero dollars, which is what you would be left with if you rejected the Donor's one-dollar offer.

In fact, however, about 15% to 20% of the time, seemingly rational people do reject a low offer, usually because they feel it was stingy or unfair or disrespectful. Just as surprising to economists, Donors on average offer about 40% of their stake. If Ultimatum is played for ten rounds, the modal offer is about 50%. My intuitive response when I was first introduced to this game and received an offer of two dollars was to reject it in a huff. My economist colleagues laughed at me, but was I irrational?

Well, in one sense my decision was irrational—*if* all that mattered is the one dollar. But was that all that mattered? In life, most exchanges are between people we know or at least may meet again at some point. In real life, if I accept a mingy offer in one transaction or another, I suffer some reputational cost—namely, that others can be disrespectful to me and I will not make a fuss. Moreover, there are costs in having others believe they can disrespect me with impunity, such as being excluded when they share food. I also incur a cost to my sense of self-worth. Consequently, my brain concludes that the person suggesting the mingy offer (Donor) should incur a cost for assuming I can be exploited without fuss. Therefore, I fussed. I rejected mingy offers.

The rejection pattern in the Ultimatum game across populations clearly shows cultural effects, suggesting that what is considered a fair offer reflects differences in cultural norms. The numbers mentioned in the previous example (over ten rounds the modal accepted offer is five dollars) are typical in the US. By contrast, in Israel and Japan, on average, slightly lower offers are accepted, and the modal accepted offer in ten rounds of the games is about four dollars out of a ten-dollar stake. In Indonesia, Mongolia, and the Amazon, very low offers, such as $1.50, were regularly accepted.[15]

In trying to understand the causes of cultural variation, the anthropologist Joseph Henrich and colleagues studied small-scale societies and found large differences among them in offer values and rejection rates.[16] Henrich suggests that part of the explanation for these cultural differences is the degree to which a group is integrated with external markets. Another part is the degree to which personal livelihood involves cooperation with people outside one's family. The point here is that when individuals are part of a larger cooperative organization, they are accustomed to sharing their surplus beyond the family boundary, and to having a wider reputation. Consequently, this background inclines them to make larger offers than would those in isolated groups. Fairness norms are like many norms guiding our behavior. They are shaped by a range of factors, including the local ecology, how the group makes a living, particular individuals who may inspire imitation, and the social style of other groups they interact with.

What is the brain's reward system doing during these transactions in the Ultimatum game? To explore this matter, neuroscientists Ting Xiang, Terry Lohrenz, and Read Montague raised the following question: Could Responders be subtly influenced in the lab to revise their fairness norms (rejection points)—either upward or downward—and could such revisions be seen in the brain? To answer this question, the researchers recorded the brain's activities while subjects were in a magnetic resonance scanner. The experimenter could pretend to be Donor and would manipulate the offers in the following manner: some Responders were led to expect rather high offers as typical, others to expect rather low offers as typical. If *norm* prediction error (normally being offered more, or less, than expected) is like *reward* prediction error, would we see revisions in norms after training? And what would we see in the VTA, the nucleus accumbens, and the frontal cortex?[17]

The 127 American subjects in the experiment can be assumed to have started with much the same rejection disposition that is typically seen in Ultimatum game transactions in the US; namely they reject offers lower than about 40% of Donor's stake. Subjects were told that on every trial, Donor has twenty dollars to disburse, and that Donor is a new person on every trial (not true, but a harmless fib needed to keep confounds out of the experiment). Each subject (Responders all) is given a total of sixty offers. The offers come in two blocks of thirty. The training consists in shaping their normative expectations by controlling the pattern of offers. There are four groups of Responders. Group L→M starts with thirty Low offers and then receives thirty Medium offers. Group H→M starts with High offers and then gets Medium offers. Groups M→L and M→H start with thirty Medium offers, then get Low or High offers, respectively.

On every three to five trials, subjects are asked to rate their feelings about their offers on a 1–9 scale of emoticons—1 being very happy, 9 being very unhappy. This rating was done to determine whether the training had any conscious effects on behavior or on how the subjects felt about how they were doing in the game.

Consider the H→M and the M→L subjects. At Trial 31, the offers began to be significantly poorer than the offers the subjects had become used to. Norm prediction errors—worse than expected—were seen in the accumbens and in the prefrontal cortex. By contrast, for L→M and M→H subjects, the offers were richer than the subjects were used to and yielded a better-than-expected response in the accumbens and the orbitofrontal cortex. These results do indeed show that the reward system responds to norm prediction error in the same way that it responds to reward prediction error.

An even more striking result regarding fairness norms came out of the data. Suppose for the first thirty trials you usually got High offers, and I usually got Low offers. Now at Trial 31, we get exactly *the same Medium offer*: nine dollars. Your accumbens signals "worse than expected," and mine signals "better than expected." Why? Because our histories, and hence our expectations, were different. You went from High to Medium; I went from Low to Medium. You will reject a Medium offer; I will accept. How do we feel about exactly the same nine-dollar offer? You are displeased; I am delighted. Moreover, our subjective ratings correspond to the activity seen in the fMRI scans of our brains.

You and I almost certainly entered the experiment with our rejection bar pegged at much the same place. Only our recent histories—our first thirty offers in the experiment—are different. After the first thirty offers (you got Highs and I got Lows), our social behavior has changed. How we felt about the very same offer reflects our norms as modified by our experience of what was normal.

Notice that the norm revision in the first thirty offers was not triggered by discussion of whether the norm was rational or defensible or proper. The norm changed because the context concerning what is normal changed: I usually got Low offers; you usually got High offers. Almost certainly I had no idea my norm had shifted in those first thirty trials. The same goes for you. (And all the other subjects in the study.)

I think this experiment is a very big deal. If the results from it bear upon norms and how they shift in individuals as context and community standards shift, then they can give us insight into everyday life. One philosophical view might claim that a norm changes only upon conscious reflection and rational choice. Although that may be true occasionally, it is not always true. The Xiang data were, of course, collected in a laboratory setting, not a street or kitchen table setting. Nevertheless, fairness norms motivating a rejection threshold are not independent of social concerns when they change. Sensitivity to our reputational value always weighs in, even when we are not consciously thinking about it, and even when we are alone in the brain scanner.

The changes in what is fashionable in clothes are a well-known case of norm shifting in the absence of rational debate. We are all familiar with the phenomenon. I am astonished when I look at 1950s photos of myself. The clothing I saw then as utterly charming now looks faintly ridiculous. Wearing saddle shoes and a pink felt skirt adorned with a poodle, I proudly fit right in. Over the intervening years, my norms regarding fashion have profoundly changed, and they did so without much—okay, without *any*—thought on my part. What is considered beautiful in fashion is not just a purely cognitive judgment; we have a significant emotional and evaluative response. The fashion industry, of course, manipulates our norms, and last year's outfit now looks, well, so last-year. Meanwhile, the midbrain dopamine system is doing its job, modifying expectations and evaluating things as desirable or aversive, as wonderful or not so wonderful after all.

For reasons that tend to be hard to nail down, community standards about many topics do change. Often the causes are very subtle, rather like the training of subjects in Xiang's experiment. We may not actually notice that a standard is slowly changing, even as our own norm shifts in synchrony. Moreover, not everyone's norms shift at the same rate. Some, given their own circumstances, may be early adopters, while others may continue to find the traditional mode preferable. In addition, you can be quite traditional with regard to some norms, and less so with regard to others.

In my lifetime, I have seen many social norms shift. Breastfeeding babies, recycling what used to be thrown out, and accepting differences in sexual orientation are cases in point. Many other norms could round out the list. Not everyone shifts in the same direction or at the same rate. Much seems to depend on where you live, your personality, your interactions outside your community, and how socially engaged you are.

THE BRAIN AND SOCIAL NORMS

The reward system is a marvel. The larger the prefrontal cortex, the richer the communicative connections to the subcortical structures. In consequence, the capacities of the reward system for learning expand in power, subtlety, and complexity. What we have learned so far about the reward system addresses these major questions: Can it play an important role in the learning of social and moral norms, complex as they are? And can it help explain the powerful feelings associated with the voice of conscience? The evidence strongly indicates that it can do both, especially because the basal ganglia are richly connected to frontal cortex and the hippocampus.

Progress continues to be made on neural mechanisms, much of it surprising and all of it complex. A great deal remains unknown, especially concerning the exact nature of the contribution of subregions of frontal cortex. One lesson to ponder in this chapter is that norms can shift in very subtle ways, depending on what is normal in our experience. Our experience of what is normal governs our expectations, which are something that the reward system works very hard to tune just right. The reward system is, after all, in the reward prediction business.

In the next chapter we will look at a neurobiological result that at first blush is astonishing, but on second thought resonates with what our wide-ranging social experiences tell us. The result concerns the interaction between personality traits and the honest verdicts of conscience.

CHAPTER 5

I'm Just That Way

From 600 B.C. to the present day, philosophers have been divided into those who wished to tighten social bonds, and those who wished to relax them. BERTRAND RUSSELL[1]

PERSONALITY AND SOCIAL ATTITUDES

Variability in what conscience judges right or wrong is an enduring fact of social life. Family dinners erupt in an uproar as kin heatedly disagree about a norm involving sexuality, such as polygamy or homosexuality or premarital sex. Feelings run high, sometimes very high. Such differences, even among those with shared parents and shared education, are common. Our everyday explanations for these differences sometimes refer to differences in personality. "Oh, Auntie Susie, she is just a natural-born rule breaker." "Well, Uncle Henry always thinks old ways are best." And we recognize certain temperamental traits in ourselves that regularly predict how our own conscience will judge—*live and let live*, or *condemn and never forgive*. Likewise, our siblings and mates respond in their predictable fashions.

As we have seen, what our own conscience requires of us depends on our instincts for sociality, but also on what we learn as we grow up in our social world. The family dinner uproar, however, invites a sideways question: Do our personality traits also feed into what our conscience finds wrong or right?

Are there differences in the brain that reflect whether we are natural-born rule breakers, or whether we favor traditional ways? Until recently, I would have said, "Not that we know of." That is not my answer now. An important neurobiological result indicates that yes, there are such differences in the brain and they can be observed using standard brain-scanning techniques, such as functional magnetic resonance imaging (fMRI). As revealed in brain scans, the brains of different people react quite differently to visual images of negative stimuli, such as a rotting animal carcass. Surprisingly, distinct patterns in brain responses to such images turn out to cluster according to social attitudes. In particular, they cluster according to whether you are likely to be traditional in your approach to social norms or inclined to be laissez-faire; or, in Bertrand Russell's terminology, according to whether you wish to tighten social bonds or to relax them. To put it yet a different way, the brain images cluster according to whether you are strongly conservative or strongly liberal in your social views. Moderates are in between.

My first run through the data fired up my skepticism, a lifelong personality feature of my own. But the more I pushed back, the more compelling I found the data. My skepticism slowly gave ground to fascination. The data are not junk neuroscience. Not by a long shot.

The rather boring title of the paper that provoked my critical reflections is "Nonpolitical Images Evoke Neural Predictors of Political Ideology."[2] Not the sort of title to grab your attention as you read down a list, but with content that is certainly touchy. The lead experimenter, Woo-Young Ahn, at the time a postdoctoral fellow in Read Montague's lab at Virginia Tech Carilion School of Medicine, juxtaposed two seemingly unrelated things: nonpolitical images and political ideology. In brief, here is an outline of the experiment.

Each of the eighty-three human volunteers viewed images presented one by one while they were in a brain scanner.[3] Twenty images were neutral, such as a mountain with a stream; twenty were negative, such as decomposing human bodies or worms crawling out of a person's mouth; twenty were threatening, such as a bear about to attack the viewer; and twenty were pleasant, such as children playing at the beach. None of the eighty images was explicitly ideological or explicitly relevant to sexuality, strong leadership, outsiders, or other hotbutton issues.

After exiting the scanner, the subjects were asked to rate how they felt about each image on a 1–9 scale. They then answered a questionnaire known as the Wilson-Patterson Attitude Inventory, a well-validated assay that rates the degree to which a subject is strongly conservative versus strongly liberal in attitudes toward specific normative issues such as authoritative leadership, foreign aid, the death penalty, immigration, and premarital sex.[4] A second test was more

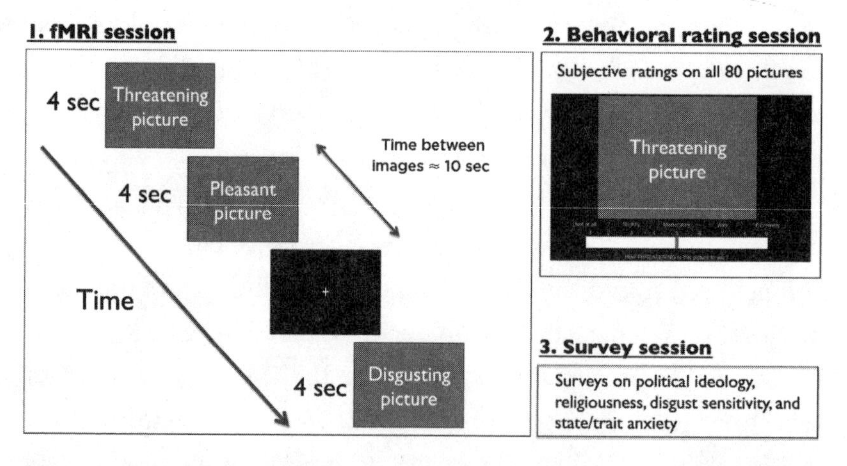

Figure 5.1 The three components of Ahn's experiment. (1) In the brain-scanning session (fMRI), subjects saw a series of pictures, each shown for 4 seconds; the time interval between pictures was about 10 seconds. Occasionally, a blank (neutral) screen was presented, and subjects were instructed to press a button as soon as they saw the figure of a cross. These neutral images were included to ensure that subjects were fully awake and responsive. (2) Subjects were asked to rate on a 1–9 scale each of the eighty pictures presented in the scanner. Subjects did not know when they saw pictures in the scanner that they would later be asked to verbally evaluate them. (3) Subjects filled out the Wilson-Patterson Attitude Inventory, as well as other attitude surveys. WOO-YOUNG AHN ET AL., "NONPOLITICAL IMAGES EVOKE NEURAL PREDICTORS OF POLITICAL IDEOLOGY," *CURRENT BIOLOGY* 24, NO. 22 (2014): 2693–99. WITH PERMISSION.

general, probing for attitudes toward bedrock principles that are less specific to a particular historical time period. These questions concerned matters such as compromising versus adhering to principles, mercy versus strict punishment, and responding first to the needs of those in one's own group versus responding to the urgent needs of an outsider (Figure 5.1).

This is a complex and careful experiment requiring thorough examination of the details, but the results were unambiguous and needed no statistical massaging. As the authors say, "Remarkably, brain responses to a single disgusting stimulus were sufficient to make accurate predictions about an individual subject's political ideology."[5]

Let me explain further. If, when you viewed the image of worms in a mouth (Figure 5.2), your brain showed *high* levels of activity in areas associated with valuation assignments, emotion processing, and preparation for action (among others), then you were also highly

Figure 5.2 The image used in Ahn's experiment was like this, but in color. COURTESY JOHN HIBBING.

likely to score on the conservative end of the Wilson-Patterson inventory. Contrariwise, if your brain did *not* react as strongly in those same brain areas, you were highly likely to score on the progressive side of the Wilson-Patterson inventory (Figure 5.3). Accuracy was about 85% in predicting whether subjects scored as conservative, moderate, or liberal on the Wilson-Patterson inventory, based on the brain response to only one negative image, such as worms in the mouth.

One surprising result was that there was little correlation between the brain response to a highly negative image and how negatively the subject rated the image on the 1–9 scale. That is, a traditionalist's brain might reveal a quite strong reaction to the worms image, even though he reported not experiencing strong feelings to that image; a progressive's brain might reveal a lower level of reaction even when he reported finding the worm image extremely upsetting. Subjectively, I did not find the worm image particularly negative or upsetting, but I have to recognize that my brain may be highly reactive to that image nonetheless. I do not know from introspection what my brain is doing.

The connection between a person's political attitudes, and his or her brain's responses to a negative image such as a mouth full of

Figure 5.3 Facing page: These gray-level images show the results from experiments in which subjects viewed images while in a brain scanner. BOLD means "blood oxygen level–dependent." Functional magnetic resonance imaging (fMRI) is not a direct measure of neural activity, but registers blood flow changes, which provide a good approximation of neural activity. (A) Response to the first disgusting stimulus for the liberal and conservative groups, a composite computed from responses in each of the brain regions shown in part B. The *x*-axis is time following stimulus presentation (in seconds), and the *y*-axis is the percent change in the BOLD signal. In this instance the conservatives show a greater brain response than the liberals. Shaded regions indicate the variation between subjects in the group. (B) Response to the first disgusting stimulus, extracted from each predictive region. Down-pointing arrowheads indicate the stimulus onset. AMYG/HIPP, amygdala/hippocampus; BG, basal ganglia; DLPFC, dorsolateral prefrontal cortex; FFG, fusiform gyrus; MTG/STG, middle/superior temporal gyrus; PAG, periaqueductal gray; pre-SMA, pre-supplementary motor area. SIMPLIFIED FROM WOO-YOUNG AHN ET AL., "NONPOLITICAL IMAGES EVOKE NEURAL PREDICTORS OF POLITICAL IDEOLOGY," *CURRENT BIOLOGY* 24, NO. 22 (2014): 2693–99. WITH PERMISSION.

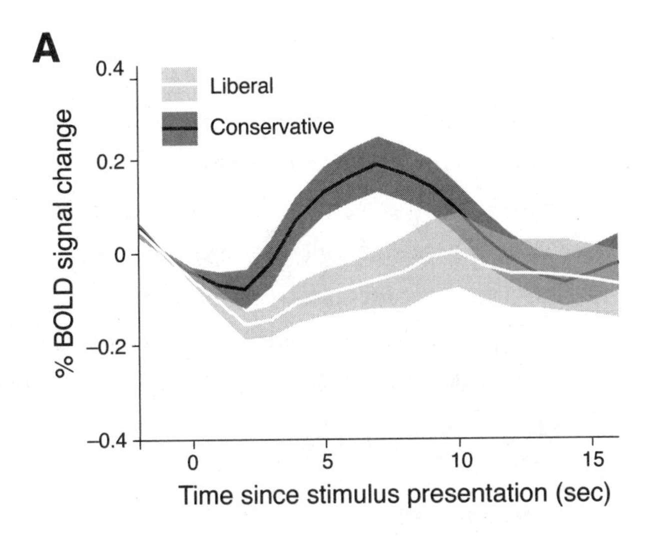
A
Liberal
Conservative
% BOLD signal change
0.4
0.2
0
–0.2
–0.4
0
5
10
15
Time since stimulus presentation (sec)

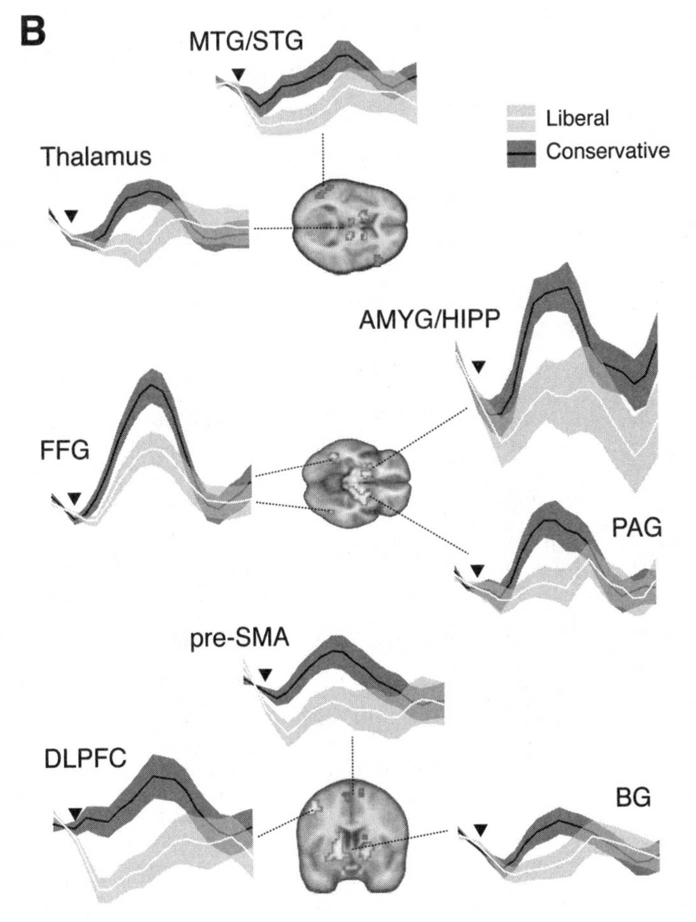
B
MTG/STG
Liberal
Conservative
Thalamus
AMYG/HIPP
FFG
PAG
pre-SMA
DLPFC
BG

squirming worms, is anything but obvious. The idea of testing for such a connection originated with John Hibbing, a political scientist, who had a hunch that such a connection might exist. Hibbing's curiosity was fueled by a range of experiments indicating differences between conservatives and liberals in behavioral responses to non-political stimuli, especially regarding those of a negative nature. For example, when conservatives and liberals are presented with exactly the same visual images, on average liberals rate negative stimuli less negatively than conservatives do. When both groups are shown emotionally ambiguous faces, conservatives tend to see anger, while liberals tend to see surprise.

In other experiments using eye trackers, when subjects viewed images—including neutral and positive images along with negative images, such as vomit, a house on fire, or dangerous animals—conservatives focused more quickly on the negative images, looked longer at them, and were inclined to fixate on them.[6] As Hibbing realized, behavioral experiments of this kind make you wonder whether brain differences observable with a brain scanner might correlate with behavioral differences. The results of the brain scans in Ahn's experiment indicated that Hibbing's hunch was right.

Another surprise in the data, however, was the very broad collection of brain regions whose activity correlates strongly with political attitudes—high activity for conservative and lower activity for liberals. Certainly, no one expected the data to reveal a single *center* for political attitudes, but the range of regions involved does not have any known unifying function. The region called supplementary motor area (SMA) is known to be involved in preparation for an action. The dorsolateral prefrontal cortex, or DLPFC, is involved in working memory, valuation updates, inhibition of inappropriate ideas, and so forth. The periaqueductal gray (PAG), is not well understood, save that it is involved in control of pain sensations. A motley crew.

Notice that the results of Ahn's experiment do not speak to causality. For all we can tell from these data, the link between the brain response and the political attitudes may depend largely on experience and learning, or it may have a significant genetic component, or both. Political attitudes and brain reactivity to a negative image may have a common cause, or perhaps the political attitude causes the characteristic brain response. Or it is possible that the degree of reactivity is part of an elemental personality feature that is causally relevant to political attitudes and perhaps other nonpolitical attitudes as well.

MY GENES AND I

What *are* relevant to the question of causality, and hence to the imaging experiment, are studies from behavioral genetics that ascertain whether traits, including political attitudes, might be heritable to some degree.[7] And indeed, the data show a *significant* degree of heritability for political attitudes. The results, reported by political scientists John Alford, Kendall Funk, and John Hibbing, were based on large-scale studies of twins, both *identical* and *fraternal* (monozygotic and dizygotic, respectively). Monozygotic (MZ) twins are genetically identical (sharing 100% of their genes), whereas dizygotic (DZ), or fraternal, twins are related as any two siblings are, and hence are assumed to share only 50% of their genes.

The crucial point is to compare MZ twins with DZ twins to see how strongly a trait is matched in the groups of twins. This is a way of finding out whether genes play a significant role in some traits, or if the environment is largely responsible. For example, MZ twins, even if reared separately, have much the same height, whereas DZ twins, whether reared apart or together, have a weaker correlation in height. One DZ twin may be short and the other tall. They are not, after all, genetically closer to each other than they are to their other siblings.

The main point is easiest to see in the legendary Minnesota studies of twins reared apart. Thomas Bouchard and colleagues analyzed fifty-four pairs of MZ twins and forty-six pairs of DZ twins.[8] For certain personality traits, including aggressiveness, traditionalism, and obedience to authority, they found that MZ twins reared apart matched up about as much as those reared together. This result implies that the environment does not make a big difference in the manifestation of these traits, but instead that they are strongly influenced by the genes.

Twins studies are not confined to twins reared apart to estimate heritability of a trait. In the more inclusive twins studies, researchers typically draw on data from very large numbers of twins. To estimate heritability, the reasoning goes like this: if in a very large population, the correlation between MZ twins for a given trait is at least twice as large as that of DZ twins, the similarity in MZ twins is likely due entirely to genetic factors. Eye color, for example, is highly correlated among all MZ twins, but not as highly correlated among DZ twins, where it is not uncommon for one twin to have green eyes and the other brown eyes.

If the magnitude of the correlation is the same for MZ and DZ twins, the similarity is likely owed to shared environmental effects. This might be true for, say, preferring surfing to playing tennis. If the magnitude of the correlation in DZ twins is between 50% and 100% of that seen in MZ twins, then it is likely that both genetic and environmental effects are involved.[9] Agreeableness is one trait where the genes are definitely significant, but the environment may also have a role. This contrast between populations of MZ and DZ twins concerning the genetic influence on a particular trait is called the *heritability* of that trait.

More exactly, heritability is a measure of how much of the difference between two populations is accounted for by genetic variation. Height is highly heritable. About 60% to 80% of the difference in height between individuals is accounted for by differences in genes;

about 20% to 40% is accounted for by environmental differences such as nutrition. What about the heritability of a disorder such as schizophrenia? About 80% of the difference between those afflicted and those not afflicted is accounted for by differences in genes. And the heritability of political attitudes? They, too, it turns out, are significantly heritable—about 40% to 50%. Let's look a little more closely at the data.

Owing to the foresight of psychologists, very large databases on twins exist—in the US, Australia, Canada, Sweden, Israel, Finland, Denmark, Japan, and the UK. The Swedish database, for example, was launched some fifty years ago and has records on more than 100,000 twins. The *Virginia 30K* database has about 14,000 adult twins, but also responses from spouses, parents, siblings, and adult children as an independent check on the twins' responses, totaling about 30,000 entries. These kinds of numbers are sufficient to yield a meaningful result.

A range of personality traits, including *extroversion*, *openness to experience*, *emotional stability*, and *agreeableness*, have been incorporated in the database and hence can be analyzed for heritability. These databases are an invaluable resource for behavioral genetics, especially because they make possible the comparison of heritability results from one culture with those of other cultures.

If MZ twins' scores highly correlate for each of these personality traits, but DZ twins' scores do not, the environment probably has rather little impact on the expression of these traits; the genes have a big effect. And this turns out to be what the results show for *extroversion*, *openness*, *agreeableness*, *conscientiousness*, and *emotional stability* (sometimes referred to as *neuroticism*). In its heritability, openness to new experiences, for example, is a little bit like height. Just as height can be affected by environmental factors such as prenatal and infant nutrition, the genes nevertheless play a major role. Likewise, openness can be affected by major life events, but it is still highly heritable.

By and large, the postnatal environment modestly affects the aforementioned personality traits, which tend to be largely stable throughout life. At the school reunion forty years on, the class clown is still cracking jokes, the agreeable activity chairman is jollying everyone into the bar, the conscientious ones are cleaning up the beer bottles. The contribution of the genes, whatever those genes are, is significant.

Other traits—for example, belonging to a particular church (Methodist or Episcopalian) or favoring a baseball team for a lifetime (Yankees or Mets)—are not highly correlated in MZ twins reared apart. They are, however, correlated in MZ as well as DZ twins reared *together*. This finding implies a strong environmental effect in preferences for church affiliation and a baseball team. If your dad always took you to the Yankees game and cheered the team on, you are likely to get into the Yankees spirit as well. Being a Yankees fan rather than a Mets fan, or a Methodist rather than an Episcopalian, is not genetically influenced.

Political attitudes, as measured by tools such as the Wilson-Patterson Attitude Inventory, are significantly heritable—roughly 40% to 50%. For example, MZ twins, whether reared together or apart, score similarly on attitudes toward gay rights, school prayer, and abortion.[10] How do personality traits link to political attitudes? Scoring high on *openness* tends to correlate quite strongly with being less traditional and more liberal, while scoring low on *openness* correlates with being more traditional and conservative. Since variability in openness persists in the human population over time, it is likely that under somewhat different conditions, each personality type is distinctively adaptive. Consequently, neither high nor low scores on *openness* should be valued as the best or ideal way to be. It depends.[11]

One caution that should be accorded the status of a mantra is this: the simple idea that there is a single gene for liberalism or conservatism is misguided, not least because even for a very highly heritable trait such as height, hundreds of genes are involved, and each plays a small role. For heritable personality traits, we can reasonably assume that, at a minimum, many hundreds of genes are involved. One consequence of this fact is that traits fall on a spectrum. They are not all-or-nothing. In addition, the environment does have a significant effect on political attitudes: roughly 50% of the differences are attributable to environment. This may not be everything, but it is not nothing. The money point is straightforward: political attitudes turn out not to be as independent of our genes as we might have assumed, but our genes do not rigidly determine our attitudes.

A semantic problem is that though our customary labels *liberal* and *conservative* may capture something of whatever brain disposition it is that our genes underwrite, they probably do not begin to do justice to the actual nature of that brain disposition. The range of brain regions showing variability in reactivity to a mouthful of worms, for example, reveals no common theme that I recognize or understand. That very puzzle is part of the reason I find the data so absorbing.

Disgust is widely assumed to be an evolved emotional/visceral response that causes humans to avoid toxins and pathogens that might otherwise kill or sicken them. Such pathogens would include those rampant in rotting animal carcasses and, perhaps, healthy worms. Observing that disgust is sometimes also a response by groups to foreigners, some social psychologists opine that the pathogen-avoidance hypothesis explains humans' hostile responses to immigration and to out-groups more generally. According to the hypothesis, an instinctive pathogen stress response to members of an out-group was favored by evolution because in Stone Age times, out-group humans often carried infections. This extension of a pathogen-avoidance instinct from rotting carcasses to foreigners was further stretched to include norm violators *within* the group whose behavior could destabilize the community. For the scofflaw extrapolation to be even modestly plausible, it is assumed that destabilization is a kind of contamination, though not an outright pathogen.[12]

Although the idea that pathogen avoidance evolved to include foreigners and scofflaws may have some merit, scrutiny invites skepticism. Recall that when subjects see disgusting images, the brain areas that strongly respond are widespread and include regions such as the supplementary motor area and the periaqueductal gray. These regions are well outside those known to be responsive to disgust associated with food.[13] A more serious problem with the hypothesis is that the one brain area known to be regularly activated by disgust to food contamination images—namely, the anterior insula of the cortex—was *not* differentially activated by the negative images in Ahn's experiment.

If our allegedly automatic disgust of foreigners is, as proposed, a product of evolution, one would predict first encounters with foreigners to be typically hostile. Unfortunately for the proposal, the facts do not cooperate. For example, the indigenous peoples of the Americas and Pacific islands were by no means uniformly hostile to Europeans first arriving on their shores. In many instances the locals were curious, and sometimes they even welcomed the strangers. One famous case is Captain Cook's 1778 arrival in the Hawaiian Islands, where the inhabitants were welcoming. The absence of hostility in such instances fails to fit with the idea that we all have a pathogen stress instinct that kicks in when we meet foreigners.

Owing to the importance of avoiding inbreeding and the benefits of trade, contact between groups of our Stone Age ancestors seems to have been quite common, so it is a bit hard to see how a genetic story for instinctual avoidance of strangers is supposed to make evolutionary sense. In any case, when out-group hostility is present, it may involve emotions such as fear and anxiety, perhaps rooted in previous experiences, rather than in disgust. Contamination disgust may be at most a tiny fraction of the explanation for hostility to foreigners and scofflaws. To be taken seriously, the additional claim that there is a genetic basis for hostility to newcomers needs at least a modicum of evidence.

Finally, because nonhuman animals such as chimpanzees, monkeys, and wolves may sometimes be wary of group intruders and strangers, out-group hostility, when it does occur, needs a more comprehensive explanation than the extension of the pathogen stress response. What we would like is an empirically grounded and compelling way of connecting brain-imaging data to the heritability data from behavioral genetics and molecular genetics, and to the behavioral data from psychologists. The emphasis here is on *data*; that is, we need good, hard evidence as opposed to just-so stories and ornamented hunches. As the adage says, without data you are just another person with an opinion.

MY CONSCIENCE AND MY PERSONALITY

Enduring personality traits, such as the capacity for openness to new experiences, emotional stability, agreeableness, and conscientiousness are probably involved in the ease of acquisition of certain norms and the strength of adherence to those norms, and hence to what someone's conscience deems right or wrong. Our brains' reactivity to images of decomposing carcasses or worms squirming in a mouth reflects a background, highly general disposition that affects, among other things, whether we find a certain social norm easy or difficult to adopt. It affects whether applying that norm fits with our emotional valence toward it, and whether we are willing, or not, to modify certain norms, such as those concerning sexuality, punishment, and interactions with out-group members.

Although research is probing the underlying brain configurations that regulate our tendencies to be agreeable or disagreeable, or to be open to new experiences as opposed to being rather apprehensive about them, making neurobiological progress on these matters is difficult. We understand next to nothing about neural mechanisms in this domain, or even whether our vocabulary for talking about these matters is approximately faithful to what the brain is doing.

John Hibbing and his colleagues make the astute observation that even though economic issues, such as free trade, business regulation, and tax policies are likely to have the greatest impact on people's lives, social issues tend to arouse stronger feelings. For example, topics concerning sexuality, punishment, aiding others, and out-group interactions evoke the strongest emotions and provoke the most ardent reactions. Those who are quick to outrage in disagreements over the permissibility of gay marriage may be tepid on the question of bank regulation or free trade, even though the first is remote from their own well-being and the second has an unambiguous impact. The issues that trigger our greatest emotional tumult are more likely connected to traits that are significantly heritable.

It is common knowledge that a range of factors may influence moral attitudes—factors such as age, education, and life experiences. Nevertheless, there are undoubtedly components that we neither routinely recognize nor consciously acknowledge. From the Montague lab, we now have evidence that the brain's level of reactivity to images of contaminating scenes is significantly related to our social judgments. Perplexing as this may be, disturbing as this may be, I find myself coming back to the basic point: the data are the data, whether we like them or not. And we cannot just make up nicer data and try to pass them off as truth.

In addition to our background temperamental dispositions, there is a domain of feelings that figure in the judgments of conscience. These feelings are associated with our varied experience in the social world. At all stages of social learning, the brain's reward system interlaces cognitive discernment with feelings that jointly shape our social decisions.[14] Learning to navigate the social world we find ourselves in, responding to approval and disapproval, discovering how to belong to the group and make our way socially—these are essential features of growing up as a social mammal. The social skills and normative habits we slowly acquire are a central part of the story of who we are, and of how the conscience functions.

If my conscience depends on how my brain is organized, are brains sometimes wired in an atypical way with the result that either we do not care at all, or perhaps we care too much? That is the topic for the next chapter.

CHAPTER 6

Conscience and Its Anomalies

Common sense in uncommon degree is what the world calls wisdom. SAMUEL TAYLOR COLERIDGE[1]

PSYCHOPATHY AND ABSENT CONSCIENCE

A decade after my book *Neurophilosophy*[2] was published, I received a phone call from an Ontario man who was deeply distressed and seeking answers. His worry was focused on his son, a child adopted in infancy and included as part of the larger family to make up four children. The lad would be the youngest, and he was warmly embraced by his new siblings. The motivation for the adoption was a strong feeling, shared by the man and his wife, that they could give an unwanted child a loving home where he would be loved and would have the same opportunities in education, sports, the arts, and so forth that their biological children enjoyed. In every way, they made him one of their own beloved brood. By the time he was four, however, they discovered that he was given to injuring their family cats and dog, and more generally, he was frequently cruel to their biological children. He also victimized neighborhood children, often in shockingly creative ways that implied considerable planning.

Applying both kindness and punishments such as time-outs and denial of treats, the parents were further dismayed as his behavior became ever worse, no matter what they tried. Once he was enrolled in kindergarten, the problems could not be confined to the home scene. He regularly found ways of inflicting misery on his classmates. By the time the boy was eleven, and having exhausted all avenues of advice and counseling, the distraught parents agreed to place him in a well-known and highly regarded residential school, hoping against hope that a different environment might bring a change that would benefit him, as well as those with whom he interacted. The new school was fully apprised of the situation, and the teachers felt that their special atmosphere emphasizing responsibility, kindness, and compassion might well shift the lad away from his penchant for causing pain and suffering to those around him. It did not work. At school, he incessantly lied, stole, and hurt others.

The last straw was the letter. For no apparent reason, the lad wrote a letter to the parents of one of the boys at school. Not a boy that was a friend or an enemy—just another boy at school. In the letter, he fabricated a story saying that the boy had gone off into the woods four days earlier and never returned, presumably having died, but that the school was hushing it all up. He signed it and dropped it in the school outbox.

The horror of what he later waved off as "just a joke" got him expelled. And that was when his father called me. Not for advice about what to do, for he and the family had explored that avenue about as thoroughly as anyone could. They had followed every piece of the mountain of advice received from psychologists and psychiatrists and teachers and preachers—advice that invariably turned out to be ineffectual.

What the Ontario man wanted to know was this: Is there something amiss with the boy's brain? Was he born that way? Does neuroscience have any answers regarding this sort of phenomenon?

Coincidentally, I had recently read *Without Conscience*, Robert Hare's classic discussion of psychopaths.[3] The book had given me a limited understanding of what was known, but certainly I was no expert. I did know, too, from my own experience growing up, that a child may be socially problematic, however loving the family and wider environment are. Still, I could not just say that and hang up. Tentatively, and acutely aware of my appalling ignorance, I suggested that it was entirely possible the lad had been born with atypical social dispositions. If these traits persisted into adulthood, he could well be diagnosed as a psychopath. I recommended Hare's book. Not surprisingly, my take on the boy's story was not news to the father. The lad's insensitivity to all forms of approval and disapproval, even as a youngster, suggested that from the get-go, he had utterly lacked feelings of remorse or guilt or shame.

Nevertheless, neuroscience was not at a stage where I could say anything at all about what might be different about the brains of psychopaths relative to more typical social humans. Is neuroscience now at the stage where more can be said? A little more, but not as much as one might have hoped. Although scientists anticipated that brain-imaging techniques might reveal the relevant differences between typical brains and psychopathic brains, the evidence based on those techniques remains tantalizing but inconclusive. Frontal structures, and their loops to the reward system and to areas important for motivation and feelings, are generally suspected as likely culprits.

The best of the brain-imaging research on psychopaths does show lower levels of activity in diverse regions—some in the frontal cortex, some in the reward system, and some in seemingly improbable places. Unlike individuals who suffered head accidents that damaged the front of the brain, psychopaths show no frank lesions or holes in images of their brains. Moreover, while some persons diagnosed as psychopaths show relatively decreased activity in frontal structures, so do some perfectly typical humans. Individual variability means that the sample studied must be very large if we are to get meaningful results.

Before going any further, we need a clarification: What is meant by *psychopath*? The behavioral criteria for a diagnosis of psychopathology are complex, entailing not only antisocial and conduct problems, but more exactly, a lack of feelings of guilt or remorse, the absence of significant bonding with others, and a lack of compassion or empathy even for those in the family who have shown them great affection.

Psychopaths are narcissistic and are pathological liars, showing no sense of embarrassment or shame when caught flat out in a barefaced lie. They are without a moral compass and can be highly manipulative, mercilessly exploiting the kindness and goodness of others. Some offenders imprisoned for ghastly murders may show conduct disorders but still be capable of a degree of remorse and shame, and they may be strongly bonded to certain family members. They may have some traits in common with psychopaths, but they are not psychopaths. The diagnosis thus involves two prongs: antisocial conduct *and* absence of appropriate emotional responses such as guilt and remorse. And of course, psychopathic traits come in degrees.

At the University of British Columbia, Robert Hare launched the research framework for the first genuinely systematic studies of psychopathy, motivated by his early study of prisoners in a British Columbia penitentiary.[4] After being conned a few times by very slick deceivers, Hare came to realize that some men in the prison were exceptional in their lack of remorse, their skill in deception and manipulation, their narcissism, and their habitual cold-blooded lying. Hare recognized that until rigorous criteria were not only proposed but generally adopted, research on the phenomenon would be hamstrung by semantic confusion and experimental confounds. He and his lab set to work devising a tool to standardize research on this special class of individuals. They called the tool the "Checklist for Psychopathy." Because they knew that psychopaths were only a subset of very nasty prisoners, Hare's lab designed the criteria to distinguish genuine psychopaths from the rest of the criminal crowd. Not every murderer is a psychopath; not every check forger, drug dealer, or habitual liar is a psychopath.

Now known universally as the "Hare Checklist for Psychopathy," this document is the current gold standard for diagnosing psychopathy. Because psychopaths are typically accomplished liars, it was obvious to Hare that you cannot assume honest answers on a questionnaire. As the American psychiatrist Hervey Cleckley had astutely pointed out twenty years before, a psychopath can be an almost perfect mimic of people who do have normal emotions. They tell convincing tall tales, with flair and a perfect mask of honesty. Consequently, the Hare Checklist requires independent background verification by parents, teachers, counselors, local police, siblings, and so forth. Administering the checklist also requires some training, since the novice is apt to assign high scores without doing a sufficiently sensitive accounting of all aspects of the subject's life.[5]

Twenty traits are assessed by the Psychopathy Checklist-Revised, where each item scores 0, 1, or 2, and a score of 30 out of a possible 40 is diagnostic of psychopathy.[6] Subjects are assessed for features like their emotional reactions following social cruelties, truthfulness, attachments to others (such as their own children), narcissism, as well as whether they feel guilt and remorse after injuring someone and whether they have a tendency to persuade others with high-flying plans and tall tales.

Does *sociopath* mean something different from *psychopath*? In popular culture, the two are often used interchangeably, but scientists prefer not to equate the two terms. *Sociopath* seems to have been coined and favored by those who predicted that the causes of socially problematic behavior are entirely environmental or sociological, with no genetic contribution. Accordingly, they chose to build that prediction into the very term itself. Moreover, they applied that term to anyone who frequently exhibits antisocial behavior, thus broadening its scope. But a grandiose and deceitful used-car salesman or a Peeping Tom or a hard-hearted foreclosure agent may not score anything close to 30 out of 40 on the Psychopathy Checklist, however obnoxious their acquaintances may find them. *Psychopath* is a term best restricted to those who score 30 or higher on the Hare Checklist. Otherwise, diagnosis becomes haphazard and meaningless. *Antisocial personality disorder* is a label that may cover some of the other bad customers.

Struggling to find a suitable label for these unusual persons, doctors in the nineteenth century used the expression "moral insanity."[7] Their observations told them that the persons in question were not otherwise delusional or out of touch with reality. In contrast to some schizophrenics, these individuals never suppose themselves to be Jesus or God; they do not hear voices or see prophecies in common objects. Nor are they depressed or unintelligent or morbidly anxious. Rather, their singular and stunning characteristic is that they lack a conscience and act accordingly. Within the domain of social behavior, they are prone to act in immoral ways, often quite gratuitously, without any benefit for themselves, and with calm planning and intent. And also with a complete lack of remorse, embarrassment, or shame. The merit of the term *moral insanity* is that it does rather capture the core of the disorder. The shortcoming is that it fostered the informal inference that psychopaths were "psycho" and hence delusional like schizophrenics, or anxiety-ridden like those with obsessive-compulsive disorders. And they absolutely are not.

Within the prison population, psychopaths account for about 25%. They do time for murder, rape, arson, identity theft, assault, and battery. They are highly likely to reoffend. Since they don't much care whether what they are doing is seen to be wrong, they are often apprehended and end up in prison. Estimating the number of psychopaths in the wider population is more difficult than for those in lockup. While bored prisoners may be happily available for psychologists to diagnose, an advertisement recruiting subjects for a psychopath study is unlikely to draw in psychopaths who are at large. For one thing, they are not in the least unhappy with themselves and see no reason to be assessed. They do not feel anything is wrong with them. All we can really do is make an educated guess about numbers. Hare and neuropsychologist Kent Kiehl concur that probably less than 1% of the population as a whole would score 30 or higher on the Hare Checklist. Greater estimates, such as 4%, are based on looser criteria and hence include disagreeable folks who, however, are not genuine psychopaths as measured by the Hare Checklist.[8]

It is also important to note that psychopathy is seen in females as well as males. Because most psychopathic serial killers are male, and because such killers are spectacularly newsworthy, there is a common belief that most psychopaths are male. But what do the data show? In the female prison population, about 25% score as psychopaths, just as among male inmates. Notice though, that, on the whole, there are about ten times as many male offenders as female offenders. Arithmetic thus implies that female psychopaths are fewer in number than male psychopaths.[9] On the other hand, recent research investigating callous and unemotional (CU) adolescents has found that the number of CU males and females is about the same.[10] Clearly, additional research is needed. Owing to a semantic convention adopted by researchers, only adults are diagnosed as psychopaths. Individuals under twenty-one who have conduct disorders and who lack remorse, guilt, and shame, are diagnosed as CU.

So far, we have focused on behavioral criteria for psychopathy, but what about the psychopathic brain? A reliable biomarker is a reduced startle response. Here is what I mean: Typically, after people are shown distressing images such as that of a charging bear, their startle response is more intense than when they are first shown neutral images. We all likely know this about ourselves.[11] We are more jumpy for a time after a threat, whether it is a near accident on the freeway or a menacing phone call in the middle of the night. Psychopaths generally do not show this enhanced startle response. What, neurobiologically, might this difference have to do with their lack of conscience?

According to one early hypothesis that tried to make sense of this atypical startle response, the amygdala, a subcortical structure known to be involved in fear responses, may be dysfunctional. This hypothesis was also supported by the common observation that punishment tends to have no effect at all on modifying the antisocial behavior of psychopaths. Punishment and the threat of punishment are frightening to normal humans. Not caring about punishment suggests dysfunctional fear processing. Maybe the amygdala is not functioning properly.

Kent Kiehl, Robert Hare, and their colleagues were the first to scan the brains of psychopaths using functional magnetic resonance imaging (fMRI) and to compare the results to those of typical control subjects. Through heroic efforts, they were able to scan male prisoners who had been transported from a maximum-security facility in British Columbia to the scanning facility in a Vancouver hospital. In their pioneering study, they did discover diminished amygdala activity in the brains of psychopaths during tasks designed to engage emotions such as fear and anxiety. But the amygdala was not the only location that stood out. Psychopaths' brains showed reduced activity also in the hippocampus (essential for spatial navigation and for remembering individual events in one's life) and parts of the reward system (e.g., the nucleus accumbens). For starters.[12]

Subsequent brain-imaging studies also showed distributed regions—some cortical, some not—where activity in psychopathic brains was somewhat different from the pattern typical of controls.[13] And to complicate matters further, one study found that activity in a subregion of the amygdala showed a relatively *higher*, not lower, level of activity in psychopaths.[14]

In a powerful brain-imaging study of 150 male criminal offenders matched for IQ, age, and other traits, Jean Decety and colleagues compared those with high psychopathy scores (above 30) to those with low scores (20 or less). They scanned the brains while subjects watched morally significant content in 148 videos.[15] A video might show a recipient being harmed (e.g., hair being pulled) or, by contrast, being helped up off the floor. After viewing the scenario, subjects were tested to see how they determined the emotional consequences of the event viewed.

In brief, the high-scoring psychopaths were *better* at identifying the emotions that were felt by the *recipient* in both harming and helping scenarios than were the lower-scoring subjects. By contrast, the high scorers were worse at identifying the expected emotion of the *perpetrator* in a harming video. Activity levels decreased in areas important for empathic concern when a psychopath took the perspective of a *perpetrator* in a harming, but increased when they took the perspective of the *victim*. Although a robust finding, this result was unpredicted, given the lack of empathic concern exhibited in the behavior of psychopaths. These data are indeed puzzling.

In addition to the brain-scan data briefly discussed here, a problem with focusing exclusively on the amygdala is that in one rare genetic disorder, the amygdala on both sides of the brain has completely degenerated.[16] Patients with this disorder have some rather subtle deficits in fear conditioning, for example, but they do not behave in the social domain as psychopaths do.

An updated hypothesis coming together from the brain-imaging work of many labs is that psychopaths have a dysfunction in a rather widespread and complex network involving both cortical and subcortical components. This broad dysfunction might account for the cluster of traits that epitomize psychopaths. The function of the hypothesized network? Among other things, "to integrate emotions into higher order cognitive business."[17] I am not sure what exactly this means.

Certainly, this hypothesis has more merit than a busted-amygdala theory. On the other hand, it is not particularly enlightening, partly because the hypothesis is more or less a restatement of the behavioral description of the pathology. That is, there is something wrong with the emotional responses of psychopaths. Rest assured, I am not so much faulting those who advance the hypothesis as I am commenting on where the science is right now. We do not really understand what relatively diminished activity in the tagged regions means in terms of functions such as attention, emotional responses, problem solving, planning, and predicting consequences. Even worse, the *normal* functions of those tagged regions are not well understood.

Despite the difficulty of framing a meaningful hypothesis, the brain-scanning data are invaluable in providing a preliminary framework. Although brain scans can reveal a lot about regions that show heightened or diminished activity, they are not sensitive to what is going on at the micro level.[18] They cannot image neurons and what they are up to. Unusual receptor distributions, for example, or cell-type disorganization, cannot be seen with brain-imaging techniques. Yet the micro level is probably where the basic abnormalities of psychopaths lie.

This possibility can be explored indirectly through genetic analysis, and atypical coding in genes for oxytocin receptors has been seen in psychopaths.[19] At this stage of science, microdetails in human brains can generally be revealed only at autopsy. Nevertheless, at least the scanning and genetic data tell us where to start looking when a psychopath's brain comes to autopsy for examination. New techniques in neuroscience are being developed at a rather remarkable rate, and the hope is that before long we will have safe and painless ways to obtain data on microcircuitry in live humans.

Where progress has been made is on the question of the heritability of psychopathological traits; more than a hundred studies have been reported. (Recall that researchers follow the practice whereby only persons who have reached adulthood may be diagnosed as psychopaths; children and adolescents may instead be described as having callous and unemotional, or CU, traits.) Results of the twins studies indicate that CU traits in children (and psychopathy in adults) are about 50% to 80% heritable. Recall that *heritability* is a population property—meaning, in this context, that some percentage, roughly 50% to 80%, in the variability across the population can be attributable to genetic differences, leaving 20% to 50% to be accounted for by nongenetic conditions.[20] These data indicate that the CU traits are moderately to highly heritable.

The role of the genes in psychopathy aside, psychiatrists have diagnosed some individuals with *acquired* psychopathy. In these individuals, a contributing factor is lack of parental bonding, such as might be experienced by infants and toddlers in a destitute and understaffed orphanage during wartime.[21] Poor parental bonding seems to have its greatest effect on emotional development, and thus on psychopathic traits such as emotional detachment and lack of remorse and shame. A second contributing factor is severe childhood neglect and abuse. In these conditions, the unacceptable behavior is more strongly associated with antisocial conduct than with shallow affect. Those who suffered neglect and abuse as infants, and score high in antisocial behavior, may still have more normal emotional responses, assuming they had some bonding to their mothers or fathers. Although not all children who suffer severe neglect and abuse show psychopathological symptoms, some children seem to be especially vulnerable.

Those who have suffered maltreatment as children are highly likely to have structural abnormalities in areas of the brain that can be seen with brain-imaging techniques.[22] And they are at greater risk than are well-treated subjects for a range of psychiatric conditions, including bipolar disorder and schizophrenia. The list of brain regions affected by maltreatment is long and includes frontal areas and components of the reward system. These areas are believed to be especially stress sensitive, and certainly maltreatment would be accompanied by high levels of stress hormones that can affect brain development. Sensory systems may also show structural abnormalities that are believed to be associated with diminished approach responses and greater avoidance responses.

A reasonable guess would be that brain abnormalities owed to maltreatment are causally implicated in the psychopathological traits seen in a subset of abused children. Thus, you would expect that those whose scans showed such brain abnormalities would be the same individuals who were diagnosed with psychopathy. Some were, but surprisingly, a significant number were not. In their review paper, Martin Teicher and Jacqueline Samson reluctantly came to the conclusion that "maltreatment abnormalities [in brain structure] were by and large independent of the presence or absence of psychopathology."[23]

The question, then, is this: What *are* the causal connections between psychopathology and brain abnormalities resulting from maltreatment, since you can have one without the other? Teicher and Samson considered the following hypothesis: Perhaps the maltreatment abnormalities visible in the brain scanner are, in some puzzling way, adaptive, given the horrible life conditions suffered by the maltreated children. If so, other brain differences, perhaps not visible in the scanner, must distinguish between those who develop psychopathic traits and those who do not—or, to put it a different way, between the resilient and the nonresilient maltreated individuals. What those differences might be is not understood.

Traumatic brain injury is also associated with disadvantageous effects, including psychopathic traits and personality changes. The causes of traumatic brain injury are highly diverse, ranging from problems that arise during birth, to falls, sports injuries, and car accidents. Damage intentionally inflicted by other humans is also a significant contributor to childhood brain trauma. In the clinic and the morgue, children are found to have damage caused by being shaken, thrown against walls, run down by vehicles, and struck with blunt instruments. The damage to the brain is typically widespread in such injuries, and may involve internal bleeding that itself causes further brain damage and can prove fatal. In the US, in the age group from one to nineteen years, injuries are the number one cause of death, and traumatic brain injuries account for 50% of those cases.[24]

The variability in the cause of injury and in the extent of injury limits research on the causal relation between the brain injury suffered and the psychopathological traits that individuals display. Obtaining genetic data on families in which children are physically and verbally abused is usually difficult. Suffice it to say that head injuries are never good, and all too often they are followed by a tragic array of cognitive and emotional deficits.

SCRUPULOSITY AND FERVENT CONSCIENCE

According to Mae West, too much of a good thing is wonderful. Knowing which good thing she had in mind, we can see her point, at least within limits. What limits, you ask? While we all realize that such limits are not quantifiable, it is dog-at-the-door obvious when they are extravagantly crossed. Psychiatrists are familiar with the effects of boundless urges from patients who seek relief from sexual obsessions. Constant intrusive thoughts and unrelenting sexual motivation are debilitating. They impede a normal social and working life. Misery also traps the incessant hand washers and the compulsive counters (how many peas on my plate, on your plate, on his plate) and those who require perfect orderliness in their milieu, without which they cannot work or play or do much of anything. These uncontrollable compulsions interfere with getting on in life. When debilitating, they are classified as obsessive-compulsive disorders (OCD). When they are not debilitating, we call them annoying.

Moral behavior, though generally laudable, can also be extreme, excessive, and self-destructive. When moral behavior knows no limits, when it is uncontrolled, it is called *scrupulosity*, a very old word familiar throughout history to priests, pastors, and rabbis. Convinced that all life is sacred, a person may be excessively preoccupied with avoiding stepping on ants. A coercive conviction that any wealth inequality requires giving away ever more of one's goods can imperil the well-being of one's family, quite apart from interfering with the reasonable pleasures and satisfactions of life. A flawed logic says that if bringing home two stray dogs is admirable, surely bringing home twenty-two is much more so. Never taking relief in the good works achieved, some scrupulants are constantly tormented by the work left undone. Larissa MacFarquhar, in her book *Strangers Drowning*, documents human behavior at the far end of the moral-intensity continuum, referring to those compelled to be ever more virtuous as "do-gooders."[25]

Are the extreme moralists just supergood people whose example we should all aim to copy? Aristotle would have said no. He famously favored moral balance over moral enthusiasm. His "Golden Mean" (not to be confused with the Golden Rule) advises us to be neither too generous nor too stingy, but wisely charitable; neither too reckless nor too timorous, but wisely courageous; and so forth. Confucius likewise counseled balance, harmony, and common sense. Both men knew full well that such advice was not quantifiable, and that they could not precisely say when you had strayed too far in one direction or the other. They knew, too, that whatever *common sense* actually is, it would grasp well enough what they meant. And it does. Those afflicted with scrupulosity, however, are obsessed with doing ever more good, and with never being good enough. The unfortunate result is normal good deeds forgone and disruption of the lives around them.

Scrupulosity is a not uncommon subtype of OCD. It manifests as an uncontrollable rigidity concerning a moral norm or religious ritual. For example, if cleaning before praying is part of the ritual, scrupulants may repeat the cleansing many times, fearing that their ritual lacks sufficient cleanliness and their prayers may not be efficacious. Some psychiatrists conclude that religious or moral beliefs are probably not the cause of the OCD.[26] Rather, they suggest, subjects with OCD may find religious observances or moral duties a fitting target for their compulsions, egged on by fears of contamination and fears of uncertainty. Other psychiatrists report that when religious training accentuates sin and guilt, including the idea that mere naughty thoughts are the equivalent of actual naughty actions, a higher incidence of obsessional and disruptive thoughts results.[27] The doctrine that sinning in your heart is every bit as bad as sinning in the world is, if taken literally, a heavy burden to bear. Those with a tendency to OCD may be less resilient than others in their ability to bat off the thought-equals-action idea as balderdash.

Religious obsessions, such as excessive concern with sinning, sex, and confessing, have a long history. In earlier centuries, men of the church routinely warned parishioners against being excessively conscientious, as we hear from Saint Alphonsus (1696–1787), who himself suffered excessive fears of stepping on anything cross-like:

> A conscience is scrupulous when, for a frivolous reason and without rational basis, there is a frequent fear of sin even though in reality there is no sin at all. A scruple is a defective understanding of something.[28]

Ignatius of Loyola, who founded the Jesuits in 1541, wrote that he focused overmuch on sinning—a fixation that he acknowledged was probably a scruple.[29] Thus assessed, his fixation was deemed to have been the work of the devil. This attribution of scrupulosity to the devil may originate in the reasonable observation that scrupulants are not just admirably conscientious, but conscientious to a fault—a fault with personal and social costs. In any case, insight into the undesirability of the behavior is typically lacking in scrupulants, and in this respect, Loyola was an exception. The lack of insight is possibly related to the hearty approval that is normally accorded virtuous behavior.[30]

One hypothesis is that those who suffer from OCD are temperamentally intolerant of uncertainty.[31] Certainty in religious matters is inherently unattainable; hence, OCD individuals may repeatedly confess or pray, in the hope of getting a little closer to being certain of God's grace or of getting into heaven. This suggestion is at the cognitive/emotional level, and while it may be helpful, my bets are on a deeper level of explanation in terms of the brain's networks—its microcircuitry.

For those in the grip of scrupulosity but without religious belief, it is performance in some moral domain, often concerned with sex, that is inevitably reckoned to fall short in some respect. Perfection in moral behavior, including control of one's wandering sexual fantasies, can never be attained. The inevitability of this failure causes distress, including fear of punishment, and provokes a renewed effort to come up to the moral mark. The scrupulant's conscience is, in effect, a relentless taskmaster, demanding the unreasonable while incessantly berating apparent moral turpitude. If psychopaths are totally lacking in scruples, their opposites are overwhelmed by them.

When, as happened with a college acquaintance of mine, scrupulosity settles on food restrictions, the person may gradually starve himself in hopes of reaching the ever-receding acceptable level of rectitude.[32] Sadly, my acquaintance eventually tried to live only on grass. As time went on and his conscience deemed the green diet a violation of the rights of grass, he switched to pure sunlight alone. This spiraling path ended in tragedy, and he died.[33]

In the general population, the incidence of OCD is conservatively estimated at 2% to 3%, while estimates for OCD with scrupulosity vary according to local semantics. Those who are considered excessively concerned with sinning in one community may in other communities be regarded as potential saints, and hence not classified as having OCD at all. What is generally agreed upon, however, is that excessive praying, cleaning, and confessing are highly disruptive to a normal life. Priests summoned three or four times a day to hear a disturbed congregant's confession have the good sense to know that the scrupulant is not able to hold a job or properly care for family.

How is scrupulosity diagnosed? The Penn Inventory of Scrupulosity, briefly, is a set of eighteen statements.[34] Subjects respond to each statement by saying how much it applies to them, using a scale ranging from 0 (not at all) to 4 (extremely). A total score of 21 or higher is said to indicate some degree of OCD. Women and men are affected about equally. A sample statement might be, "I frequently wash or clean myself because I feel contaminated" or "I get nasty thoughts and have difficulty getting rid of them." To be sure, the threshold of 21 is a little arbitrary, but some threshold between the far extremes is clinically useful, and psychiatrists generally agree that 21 fits their clinical experience reasonably well.

Very little is known about the underlying brain differences between those afflicted with scrupulosity and those with more typical but earnest moral concerns. It is highly probable that in those with OCD, something about the reward system is not quite right. Some brain-scan results suggest that the normal handshake between the reward system and frontal cortex is problematic in some way.[35] But in exactly what way, as well as why certain useful habits, such as washing your hands or checking your heartbeat, can become uncontrollable habits, is still puzzling.

People with OCD may respond quite well to selective serotonin reuptake inhibitors (SSRIs), but then again, they may not. Serotonin is a neuromodulator normally found in widespread regions of the brain, and as we saw in Chapter 3, it has an important role in the reward system. Drugs such as Prozac, widely used to treat chronic depression, affect the serotonin system by augmenting the availability of serotonin to bind to receptors on neurons. (The *selective* part is just that the drugs target serotonin as opposed to something else, such as dopamine.) Why SSRIs have the effect they do in treating OCD or depression is not yet well understood. Nor is it understood why a person with OCD may fixate on one kind of behavior, such as religious observances, rather than another, such as dietary restrictions.

Our conscience can be badly behaved. Nevertheless, even as our moods soar and plummet, even as we are stymied and stumped by what life throws at us, we like to believe that despite everything, our conscience abides, steady as a compass. Except it isn't always steady, and even a magnetic compass can go haywire, as when you approach the Arctic Circle, for example. What helps keep us steady through our own perturbations are our social institutions and our social life in general. And when they are threatened, we are apt to feel very much adrift.

CONSCIENCE IN EQUILIBRIUM

Aristotle's view was that balance in moral judgment, as in practical judgment generally, is a skill. He thought we would all do well to cultivate that skill as effectively as possible, as a defense against social tumult, and as a way of getting through trouble. Chiming in with Aristotle, the ever-sage Dr. Seuss advises, "Step with care and great tact, and remember that Life's a Great Balancing Act."[36]

There are likely many ways that the brain can achieve the kind of balance that Aristotle had in mind, and probably each of us finds a unique way. Suppose, for a moment, that we take the balance metaphor a bit more literally and consider what it takes to maintain our body's balance. The problem of body balance must be solved by all brains, and in two-legged animals such as humans, the problem is especially demanding of the nervous system. As infants, we learn to sit upright, then to crawl, to "cruise," and finally to walk. The brain uses resources from our foundational balance mechanisms—the vestibular system—as well as feedback from muscles, tendons, and joints, and from eyes and skin. It is all about balance and not falling over. Suppose that achieving life balance is a bit like growing into efficient and comfortable two-legged locomotion.

In learning to walk efficiently on our hind legs, our brain has to deal with the particular bodily arrangements it finds itself stuck with. We toddle around on very short legs as babies but eventually settle into a typical stride, even as our bodies grow and change. We can maintain our balance whether we are short or tall, thin or plump, or have mostly slow-twitch or mostly fast-twitch muscles in our legs. The brain will cope well enough even if one leg is shorter than the other, or if we are carrying a 30-pound pack. We can use a cane or crutches, and our brains adapt quickly. We can skate and ski. As we practice, our balance is maintained. Moreover, with effort and attention we can alter the quality of our posture and hence our movements, as actors or gymnasts may be asked to do.

We depend on our vestibular system—inner ear, brainstem, and cortex—to rebalance us when we stumble. Our arms go out to break our fall, with no conscious calculation required. We depend on the vestibular system to know the difference between self-motion and being moved. Without our conscious effort, our vestibular system keeps us in balance. We know which way is up.

Aristotelian balance, as we make our way in life, may be a bit like that. Our brains have to manage with the temperaments, personalities, and energy levels we happen to have, with our particular emotional mix and preferences and goals, and with the trials and tribulations that life throws at us. Sometimes we are brought low by tragedy or hoisted perilously high by success. We may trip up accordingly. Usually, if we are well-grounded, we ruminate and reflect on ourselves, we listen to our honest friends, and then our brains eventually find their way back to our *set point*. In other words, our brains tend to settle back into where we should be.

Why would our brains have such a set point? Because typically we are wired for sociality; we need to belong, to have friends, and to live a social life. We are also wired to be prudent and to see to our own needs. If harmony and balance are maintained over the long run, we have a better chance of surviving long enough to reproduce. Resilience, hope, and determination are biological adaptations that keep us going when times are tough. That kind of psychological balance may depend on how our neuromodulators, dopamine and serotonin, balance each other so that we stay calm and carry on (see Chapter 3).

I do not want to make too much of this analogy between vestibular balance and Aristotelian balance. Obviously, it has limitations. I am egged on, however, by noticing the range of balance metaphors commonly invoked to describe psychological balance or lack thereof. The metaphors go from describing a person as unbalanced and dizzy to unstable and unhinged. People are said to trip up, make a false step, stumble, fall off the wagon, or, with luck, straighten up and fly right. Advice is sometimes packaged with balance metaphors; for example, equanimity is balancing between craving and indifference.[37] Lots of unrelated metaphors are also common, and perhaps balance metaphors have no special status. Nevertheless, what I do find modestly useful in the analogy is its orientation to our biological nature and its evolution.

A different way of thinking about our conscience and its set points derives from the Scottish economist Adam Smith (1723–1790). He explained that when considering an action and pondering its moral status, he would go into simulation mode. He thought of himself as two persons; one was the examiner, the other the person examined. Conscience is the examiner. To be an effective examiner, Smith suggested, requires distance and impartiality:

> We endeavor to examine our own conduct as we imagine any other fair or impartial spectator would examine it.[38]

Recognizing that we unintentionally scheme to see ourselves in favorable light, Smith goes on to suggest that initially we may envision how a friend would respond and judge us. That is fine as a start, but our wily, self-serving natures may imagine rather more generosity from friends than we actually deserve. So Smith argues that for a more mature result, the simulated judge must be as close to truly impartial as we can envision. How to get that? Well, don't choose to simulate a relative or a friend or anyone who might stand to benefit from your action. Moreover, we should envision not only the impartial spectator's cognitive appraisal, but also how they would feel.

Not uncommonly, people now ask of a choice, "What would Jesus do?"—which is a reasonable way for Christians to follow Smith's advice for certain dilemmas. Smith realized that there would be variability in whom we choose to simulate as our impartial spectator. It might be Martin Luther King, it might be Winston Churchill, or a particularly frank and fair-minded teacher. It might be a person whom we actually consult, rather than just simulate, such as an attorney or a counselor.

Modeling one's behavior on admirable, if imperfect, people is a typical feature of human development. As we learn of imperfections in those we otherwise admire, we bracket the imperfections as we simulate their best judgment. That we model ourselves on others is also why it matters so much that charismatic public figures not be scoundrels or swindlers or pedophiles. And why it matters that we ourselves not be cowardly and stingy and mean-spirited. As we saw in Chapter 4, what we consider normal in the social domain can be rather malleable, shifting without our conscious awareness.

Eventually, as this method of simulating judgment of an impartial spectator becomes second nature, according to Smith, the need for effortful simulation will decline. You will have internalized impartial judgment as an unconscious part of your decision-making process. Or as I might say, the basal ganglia–frontal connectivity has settled on a stable configuration to handle most decisions. Still, the strategy can be abused, especially if we are not entirely honest in considering what, for example, the Buddha would do. I might be cheating if, in my simulation, the Buddha approves of my looting my 401(k) to gamble in Las Vegas. Smith certainly realized this all-too-human weakness, but he likely regarded his strategy as leading to a result that, at least in this world, is about as good as things get.

Smith's idea, actually adapted from Plato,[39] was his way of coming to grips with the fact that conscience is the outcome of the brains in our heads. Conscience is a brain construct rooted in our neural circuitry, not a theological entity thoughtfully parked in us by a divine being. It is not infallible, even when honestly consulted. It develops over time and is sensitive to approval and disapproval; it joins forces with reflection and imagination and can be twisted by bad habits, bad company, and a zeitgeist of narcissism. Not everyone develops a conscience (witness the psychopaths), and sometimes conscience becomes the plaything of morbid anxiety (as in scrupulants). The best we can do, given all this, is to aim for understanding how an impartial spectator might judge us. If this sounds like common sense, I suppose it is. Adam Smith and David Hume are now, as ever, shining exemplars of common sense.

No good comes of insisting that unless conscience is infallible or religion provides absolute rules, morality has nothing to anchor it and anything goes. For one thing, such a claim is false. For another thing, we do have something to anchor it—namely, our inherited neurobiology. In addition, we have the traditions that are handed down from one generation to another and, to some degree, tested by time and over varying conditions. We do have institutions that embody much wisdom. Those are the anchors. Imperfect? Yes, of course. Still, an imperfect foundation is better than a phony foundation. What we don't want to do is fabricate a myth about infallible conscience or divine laws, peddle it as fact, and then get caught out when people come to realize, as they most assuredly will, that it was all made up.

Thus a biological take on moral behavior and the conscience that guides it. Nonetheless, very different theories concerning the origin and nature of moral behavior in humans do exist and are widely believed. How does the biological perspective stand up to these other approaches? That topic is next.

CHAPTER 7

What's Love Got to Do with It?

What is most likely to be universal are those impulses that, because they are so common, scarcely need to be stated in the form of a rule.

JAMES Q. WILSON[1]

BY THE RULE

In 1989 I was invited to go to Los Angeles in response to a request from the Dalai Lama, who wished to learn some basic facts about the brain. Our little group consisted of four neuroscientists: Larry Squire, tapped to talk about remembering events and episodes; Allan Hobson, to explain what is known about sleep and dreaming; Antonio Damasio, to discuss data on the role of emotions in decision-making; and me, tasked with addressing the question of whether scientists generally believe there is a nonphysical soul in addition to the physical brain, or whether it is the brain that dreams and thinks and feels. I outlined the broad range of data that imply there is probably only the brain, but no soul.

It was a private conversation between the Dalai Lama and the four of us, with Buddhist monks assisting in translation when needed. During lunch at a long table in the garden, I began a conversation by asking the monks what in Buddhism corresponds to the Ten Commandments. (Even in retrospect, this moment remains acutely embarrassing to me, as it showed my complete ignorance of Buddhism. Still red-faced, I include it because it is essential to the story.) Respectful of my ignorance, the monks gently explained to me that morality in Buddhism is not rule based. Instead, it fosters compassion, prudence, and the habit of looking at a problem from many angles and seeking advice from elders. Like any other group, Buddhist social groups have social norms, but these norms function more as guidelines to be consulted than as rules to be rigorously followed.

In his public lecture later that evening, the Dalai Lama expanded on these core ideas about morality, and like my lunch companions, he explained that strict rules could be an impediment, whereas flexible guidelines were more apt to aid real-life decision-making that involved an individual's particular history and situation. Stories, admirable role models, and the development of strong social habits were more effective as teaching tools than was rote rehearsal of inflexible rules.[2] Listening intently, I realized that the Dalai Lama's take on morality was more in line with Aristotle and Hume than with the rule purveyors (so called for short) that so dominate much of Western moral philosophy, at least in academia. *Real* moral theory, the rule purveyors stipulate, consists in finding universal rules and showing how they originate in religion or in pure reason.[3]

Driving back to San Diego, I wondered, Do these differences between rule purveyors and wisdom seekers show up if we compare the quality of social behavior observable in their followers? Do rule-purveying societies display greater moral decency and virtue than the societies of wisdom seekers? Not that I could determine. No one could seriously claim that because their morality was not rule based, Buddhists in general were morally inferior or somehow depraved.

My cursory comparison of actual behavior across cultures prompted me to look anew at the presumptions governing the moral theorizing of my contemporaries. If a society oriented to wisdom seeking is no worse than a society oriented to strict rules, maybe *strict universal* rules are not essential or even advantageous for bringing about morally good behavior. I began to wonder, Why did most contemporary moral philosophers sideline Aristotle and Hume as unsophisticated in the moral domain? Moreover, with a few notable exceptions, contemporary moral philosophers reject outright the idea that biology has anything to teach us about the nature of human morality.[4] Why, unlike Aristotle and Hume, do they see no significant role for our biology in moral decision-making?

THREE PROBLEMS FOR THE RULE PURVEYORS

First, the rule purveyors' stipulation that morality is not morality unless it comprises universal rules is, at bottom, just a stipulation. Or, as my students impertinently ask, "Says who?"

Second, the goal of the rule purveyors in moral philosophy—to define universal and exceptionless rules that truly characterize morality—has suffered a conspicuous lack of success. Even among those who share this goal, there is no consensus that the goal has been reached, or even very nearly so, many decades of effort notwithstanding. One major drawback is that there are deep flaws even in the favored candidate theories, as we will see here. Worse, each candidate theory scorns the others as hopelessly defective. These flaws endure, despite years of clever attempts to wrestle them into submission. Oddly, for those who persist in adhering to the rule purveyor dogma, the failure of the strategy rarely provokes a reassessment of the assumptions, but instead seems to encourage ever-greater devotion to them.

Third, and most critically, in reality decision-making is typically a constraint satisfaction process. Faced with a choice, whether a physical or socially significant choice, our brains integrate many constraints and come to a decision about what should be done. Some relevant constraints are facts about the observable situation, as well as predictions about the consequences of various options; some are facts about others affected by my choice—their current and expected states of mind—and what would be socially normative in this situation; some are facts about available resources, as well as one's own capabilities and preferences. Often there are time constraints, which may limit the evaluative process. Importantly, some constraints are *values* concerning friendship, education, private property, beauty, and so forth.

Constraints will vary in their importance to the person and hence in the weight given to them. For example, some will strongly evaluate fairness, others less so. Constraints will also vary in how certain they are evaluated to be, as some predicted consequences may be deemed highly probable, while others are merely a hope and prayer. We draw on years of experience, integrating across multiple timescales; we simulate outcomes in imagination, and generalize from relevantly similar kinds of cases drawn from memory. These are the sorts of elements that figure into memorable stories that depict uneasy moral meshworks: in Homer's *Odyssey*, Shakespeare's tragedies, Ibsen's dramas, and Alice Munro's short stories, to name a small handful of examples.

This list of the three problems that rule purveyors face highlights the logic of the contrast between moral philosophers who are strict in their requirement for universally applicable rules, and those who acknowledge that, although laws undergird a formal legal system (when there is one), undergirding moral judgment as it actually functions in humans are instincts, habits, norms, social skills, values, and context-sensitive problem solving.

To flesh out the logic, we need first to ask, What are the achievements of the rule-based philosophical theories? In what follows, I will consider the three leading contenders for rule-based moral theories. In the spirit of Socrates, I will unflinchingly pit their flaws against their merits.

RELIGION, PURE REASON, AND RULES

Let's look first at the idea that human morality emerges from laws derived from religion. I touched on this idea in the Introduction, but a somewhat deeper look is appropriate now.

According to some religions, morality originates in a divine being. Thomas Aquinas (ca. 1225–1274), who favored the idea that God is perfectly rational, believed that if we have received God's grace, then we will know what we are morally obliged to do. God, perfectly rational as he is, will provide the rational directive concerning whether a particular action is right or wrong as we try to live by the Ten Commandments. Facing a moral dilemma, such as whether the commandment "Thou shalt not kill" includes animals or humans convicted of a capital crime, if we have grace we will do the rational thing, thanks to God having put us in the know. Certainty about what is right and wrong is, apparently, itself a sign of having received grace. If I am ambivalent on certain moral questions, such as capital punishment, then I know that I have not yet received grace.

While appealing in some ways, the religious approach papers over the divergences in what distinct religions regard as morally wrong. Such variability is an obstacle not just for Aquinas and his ideas, but for religious approaches more generally. One awkward problem stems from different moral judgments voiced by individuals, each claiming receipt of grace or some form of message from God. There are no compelling options for dealing with this problem because there is no independent way to check on the claim.

The next awkward problem is that even sects within one religion may disagree. In Christianity, for example, Protestants may differ from Catholics regarding the morality of using contraception and the moral authority of the pope. A favored stance used to be to declare that only *my* sect knows the real moral truths; others are regrettably misguided. Or, to put it more strongly, the others are heretics. Lacking independent evidence, and confounded by honest but delusional folks claiming to *be* God, this strategy has fallen on hard times, and theologians have mostly set it aside.

Importantly, some large-scale religions, such as Confucianism and Buddhism, have no provision at all for a divine being who dispenses moral laws for us to follow, or even a divine being of any kind. Lack of a divine being is also frequently true of so-called folk religions, such as those of many indigenous tribes of North America. As noted earlier, morality in these religions does not consist in a set of rules handed down from a deity, but in something like wisdom acquired through experience, imitation, and reflection.

Much has been written on the topic of a supernatural basis for morality, and the Socratic dialogue *Euthyphro* remains among the most powerful and decisive discussions.[5] As they are walking to court, Euthyphro confidently explains to Socrates that the gods are the source of right and wrong, giving the impression that he thinks Socrates is a bit witless to ask about the origin of morality. Socrates muses, Do the gods say that something is right because it *is* right (hence it is right on independent grounds), or is something right because the gods *say* that it is right (saying makes it so)?

Socrates has clearly laid out the dilemma: If the first option, then God is just a conveyor of moral information but is not the origin of moral information itself. If the second, then *whatever* God says is right; it is right on his say-so, however horrible or arbitrary the practice in our eyes. Euthyphro is stumped. Neither alternative is palatable as an account of the origin of morality. No other alternatives present themselves. The upshot is that deriving morality from religion does not get you what you expect, leaving wide open the question of where our moral understanding comes from.

Many secular philosophers approach the issue of the origin of moral knowledge by appealing not to a divine being, but to reason. This approach does have a connection to Aquinas, inasmuch as rationality is seen as the wellspring of moral knowledge. The difference is that God drops out of the story. In his famous 1979 essay "Ethics without Biology," the distinguished philosopher Thomas Nagel outlines the idea:

> People have, to greater or lesser degrees, a capacity for reasoning that follows *autonomous standards appropriate to the subjects in question*, rather than in slavish service to evolutionarily given instincts merely filtered through cultural forms or applied in novel environments. Such reflection, reasoning, judgment and resulting behavior seem to be *autonomous* in the sense that they involve *exercises of thought that are not themselves significantly shaped by specific evolutionarily given tendencies, but instead follow independent norms appropriate to the pursuits in question.*[6]

Morality, in this view, is independent of our neurobiology. It is constituted by what Nagel calls an autonomous domain of truths discoverable by reason. It is not even *shaped*, he avers, by instincts. Moral rules, not just customary ways of doing things that are shaped by instincts, are what reason seeks to discover. According to the view, moral rules tell us what is right and wrong—*truly* right and truly wrong—not just what some culture happens to think is right and wrong. Social practices merely tell us what the community conventions are.

To repeat, according to this approach, if rules are *proper* moral rules, they reflect "independent, stable moral truths" apprehended by our rational faculty.[7] *Independent* in this context means independent of our social instincts. (I am repeating this because I am still incredulous at the statement of the position.) The proper moral rules—the *real* ones, as it were—are universal, and hence cannot vary across cultures. Any culture that does not adhere to those rules is morally wrong. In short, the proper rules are rather like stable mathematical truths that never vary. The truth does not vary, even if cultural beliefs vary. Everywhere, 2 + 2 = 4.

From this perspective, philosophers typically claim that to grasp these moral truths requires rationality (which nonhuman animals allegedly do not have) and consciousness (which nonhuman animals allegedly also do not have). Rationality, a uniquely human capacity, enables us to disengage from how our biology bids us behave, and to apprehend universal moral truths. Or so it is claimed.

This view, concisely bundled up by Nagel, is, as far as I can tell, the mainstream view within academic philosophy, at least in the US. The German philosopher Immanuel Kant (1724–1804) is the watershed for these ideas; hence it is helpful to see what Kant contributed to moral theory that resulted in his powerful influence.

According to Kant, a choice is not genuinely moral if it is done for pleasure or joy or satisfaction, or if it is prompted by any emotion, such as compassion at the plight of the suffering or love of family or felt brotherhood. Only if it is done for the sake of duty alone can a choice be genuinely moral.[8] Though Kant appears to have believed in the Christian God, he recognized the inadequacies inherent in grounding morality in a lawgiving deity, inadequacies brilliantly exposed by Socrates. To remedy the deficit, Kant hoped he could derive moral rules, or at least a fundamental moral rule, from pure logic alone. As, perhaps, God himself does. Kant wanted to articulate a foundational principle, one that is immune from rational disagreement. He assumed that pure reason would command assent from all rational beings, so disagreement would fade away. And only pure reason, uninfluenced by emotions, can tell us where our duty lies. Kant put his vision this way:

> The ground of obligation must be looked for, not in the nature of man nor in the circumstances in the world in which he is placed, but solely *a priori* in the concepts of pure reason.[9]

This statement makes Kant's point that our nature, along with our biological inclinations to care about others and to be social, should be set aside as irrelevant to morality, often even antithetical to it. This is what makes reason *pure*. Well, then, if pure reason is going to lay down the moral law once and for all, what does pure reason deliver?

The grounding principle that Kant put forward goes like this: A rule is morally correct if and only if you can rationally endorse it as applying universally—to all and sundry, *including to yourself*. That means it applies to all people at all times and under all conditions. No special pleading, no dodges. Just the purity of logic alone. Who could object to that? Once that principle is in hand, Kant thought, our pure reason can use the strategy to deduce a set of exceptionless rules about lying and cheating and so forth and apply those rules to life's moral issues.

The Kantian program stumbles coming out of the gate. The problem starts with the phrase *rationally endorse*. Kant and his followers seem to have assumed that no one could rationally endorse a rule that was what you and I might intuitively call "unfair," for example. Because, according to Kantians, any rational person would reason as follows: if a rule was unfair to some, it could be unfair to me. No reasonable person would agree to a rule that might disadvantage him, perhaps in very serious ways. For example, if a rule allowed for enslaving people taller than 6 feet 4 inches, that rule would apply to me as well as to everyone else, and hence I would suffer were I over 6 feet 4 inches tall. Oops. According to Kantians, I cannot rationally endorse that rule. To do so would be inconsistent: I would be saying yes to the rule, then no to the rule. That is, I would be contradicting myself, which is irrational. So the rule is to be rejected by any rational person. That is how Kant's proposal is supposed to work.

The core problem is that mere logical consistency has no moral heft whatsoever. Just as arithmetic has no moral heft whatsoever. You cannot get morality out of merely not contradicting yourself.

The fatal test of the Kantian strategy is to propose a morally odious rule and ask whether someone might be consistent in endorsing the rule—that is, say yes to it, without also saying no to it. As an example, consider a real Nazi rule that children with autism should be euthanized.[10] A committed Nazi on the Kantian team might eagerly agree that this rule should apply to all people under all conditions, *even himself*, were he unfortunate enough to be born with autism. He is not contradicting himself. According to Kant therefore, the consistent Nazi does rationally endorse that rule, and thus the rule *is* a moral rule. Notice that *any* run-of-the-mill fanatic or extreme ideologue can adopt the Nazi tactic. All he has to do is be logically consistent. All he has to do is not contradict himself. Feelings of aversion or disgust, in the Kantian view, are irrelevant.

A tempting reply by the Kantians is to say, "The proponent of the euthanasia law, for example, is obviously irrational." But on what basis does the Kantian say that? "Well, because the rule proposed is morally odious." Wait one minute here. For Kantians, the whole idea is that pure rationality is supposed to tell us what *are* the moral truths. The idea collapses if we rely on our previous intuitions about what is morally correct to tell us what is and is not rational. If you smell circularity in the Kantian reply, you are entirely right. It reeks of circularity. To emphasize the point, notice that however nasty he is, the Nazi is at least not inconsistent; he is not contradicting himself. Quite the contrary. He is entirely consistent. His logic, qua pure logic, is watertight. For Kant, that is enough for the Nazi's rule to qualify as moral. Thus is the hopelessness of the Kantian program exposed.

Like our not-so-fictional Nazi, all manner of weirdos can *consistently* accept morally obnoxious rules, happily agreeing that if they themselves run afoul of the rule, so be it.

Morality does not and cannot emerge from pure logic alone. It cannot be disengaged from our deep desires to care for others and for those with whom our welfare and prosperity are entwined. It cannot be disengaged from our need to live social lives. David Hume saw this clearly. He realized as well as any other philosopher in history that it is part of our nature to have care and concern for others, and to learn the ways of our community. He observed that with maturity comes prudence in assessing the standards of community and in figuring out how to deal with life's problems. Caring, learning, and prudence are the triplet dispositions, interlocked and interdependent, that sustain a moral conscience. Kant opposed all this, placing his bets on pure reason alone.

Although the Kantian approach has found a warm welcome in many philosophy departments, it seems much less at home in the wider world.[11] Consider Kant's conviction that proper moral rules do not admit of exceptions. As even modestly clever undergraduates are quick to point out, no rule seems to be immune to fair-minded exceptions. Always tell the truth? Always, no matter what, keep our promise? Even if it means utter and certain catastrophe for all affected, even if it means the death of innocent people? Although it is not acceptable for me to lie just when it happens to suit me, there can be situations when lying prevents a large-scale disaster, or even where telling the truth is just unnecessarily rude. Moreover, there is no yet-deeper rule to specify when breaking the no-lie rule is morally acceptable.

Even the Golden Rule—"Do unto others as you would have them do unto you"—works well only when we share background values. In high school I had a boyfriend who converted to Scientology, the practice of which he regarded as overwhelmingly valuable. He wanted to follow the Golden Rule and convert me also, since if we exchanged places and he were in *my* place, that is what he would want. No thanks. Or, as George Bernard Shaw puts it, "Do not do unto others as you would that they should do unto you. Their tastes may not be the same."[12]

When I was an undergraduate, Kant's approach, especially his firm conviction that our nature and the conditions of our lives are irrelevant to morality, struck me as odd. For all Kant's apparent veneration of rationality, his story seemed no more rational than that shopped by itinerant evangelists who came through our up-country towns preaching hellfire and damnation. Just words, words, and more empty words. Too much the country bumpkin, I did not appreciate that Kant was considered a great moral philosopher.

In contemporary moral philosophy, the arch competitor to a Kantian approach is utilitarianism. Like Kant, the utilitarians sought a secular basis for morality, but unlike Kant, they thought that evaluating the consequences of a plan was more important than strictly adhering to a rule such as "Never tell a lie." As we recognized earlier, sometimes telling a lie is exactly what should be done to prevent an unmitigated catastrophe. The utilitarians rightly realized this. The founders of utilitarianism were mainly Jeremy Bentham (1748–1832), Henry Sidgwick (1838–1900), and, in more sophisticated form, J. S. Mill (1806–1873). More recent prominent utilitarians include the philosopher Peter Singer and the neuroscientist Joshua Greene.

Like Kantians, the utilitarians favor the ideal of a single fundamental rule (they call it a *principle*), which would cause disagreement to dissolve and from which all correct moral decisions could flow. The utilitarian contribution depends on the apparently uncontroversial point that happiness is better than misery, and the assumption that what matter morally are only the consequences of an action to the happiness of those affected by that action.

What exactly does utilitarianism say? The principle of utility has a beguiling simplicity: *Act so as to produce the greatest happiness for the greatest number.* Or, in a shorter formulation, *maximize aggregate utility* (by *utility* is meant things that the affected people want or, in some formulations, that contribute to their happiness).

But how do utilitarians go from the premise that each of us seeks our *own* happiness to the moral directive that each of us *should* seek the happiness of *everyone*? To take an analogy based on discussion by Simon Blackburn,[13] while it may be fine to say that each individual should floss his teeth, it does not follow that each individual should floss everyone else's teeth. Blackburn's point is that some argumentative fill is needed; some premises are missing and need to be slotted in. The fill may take the form of claiming that for individuals in general, social life is better than an isolated life—a background assumption that, by and large, each of us is better off in a population where there is more well-being than misery. The fill for the hole in the argument takes some doing, however, and utilitarians often do a fine job of pretending that the hole is not there.

There are more serious flaws. Because they adhere to the "maximize *aggregate* utility" rule, utilitarians insist that every individual in the aggregate be treated on an equal footing with every other for the purposes of calculating the utilities. No special status is to be allowed for family members or friends. This strict adherence to impartiality seems admirable at first, but it gets them into a conflict with basic instincts of social mammals such as humans.

In particular, the *impartiality requirement* puts utilitarians in opposition with such moral convictions as "Charity begins at home." It implies, for example, that my duty to provide for twenty orphans on the other side of the planet is greater than my duty to provide for my own two children. It implies that my one aging mother should count much less than five homeless people; she should count 0.2 as much. If I donate a kidney, impartiality requires that it must not be to a young sister who desperately needs it; I must donate it to the donor bank.

Some face-saving accommodations have been offered by utilitarians to soften these alarming implications. These accommodations fail for the deep biological reason that caring for those to whom we are attached is what we are wired to do and is a crucial part of what gives life meaning. Cutting that instinct out of consideration as irrelevant to moral deliberation has been criticized as itself morally indecent.[14]

The special significance of family and friends in many of our daily actions continues to challenge the principle of utility as a simple, universal rule to guide all morally significant decisions in our lives. From a biological perspective, such as the one I adopt, this feature of utilitarianism seems flatly unworkable. Most of us could not possibly live according to the demands of impartial utilitarian principles. Utilitarians, it turns out, admit that they cannot either.[15]

Why, then, insist on thorough impartiality in all moral decisions? Their answer: because it would be best if we could live that way. This is a totally unverified claim. There is indeed a place for impartiality, such as in a court of criminal law, but always and everywhere? For all my moral decisions? Stubbornness on the impartiality principle takes us perilously close to the immoral end of the spectrum.[16]

The philosopher Clark Glymour points out that a major constraint in moral theorizing is that *ought* implies *can*. What he means is that no one is obliged to do something he or she absolutely cannot do—running if you are paralyzed, for example. The Glymour test for utilitarians is this: *Can* mother and fathers care for twenty unknown orphans while neglecting their own two babies? The utilitarian math says they *ought* to. But *can* they? By and large, no. I do know that my conscience simply would not let me neglect my own two children in order to care for twenty orphans. Love for one's own family members is a colossal neurobiological and psychological fact that mere ideology cannot wish away.

Maybe the reason that utilitarians tend not to weaken the strict impartiality requirement is that without it, there is not much of a utilitarian principle left. *Maximize the greatest happiness for me and my tribe*, for example, seems a bit anemic, morally speaking. They have really nothing to put in place of strict impartiality.

Calculating how the consequences of an option affect the happiness of everyone also poses a problem for utilitarians, though this is the one place where they are supposed to shine. In reality, our assessment of happiness impact varies as a function of our background values. Some of those values concern what makes for a morally good life, and these assessments diverge across individuals. A religious hermit may have very different basic ideas about what makes for a good life than will a highly social nurse or a navy captain or a lobster fisherman or a dairy farmer. Some find a mild climate to be essential to contentment, but many adventurous souls have a grand passion for the far north, its lack of creature comforts, its wildness, its long, cold, dark winter nights. Yet the only value that is supposed to enter into the calculation concerns value-neutral happiness. This requirement, too, is highly unrealistic.

As the philosopher Owen Flanagan remarks in his insightful discussion, there is no single route to happiness, no one set of events that are conducive to a good life or to happiness for all humans. This is because humans start off life with different personalities, and our experiences in life may mean that we find satisfaction and joy and contentment in very different kinds of things. Flanagan makes the point clearly: "One problem is that the most natural environment conducive to well-being cannot be fixed for persons in the way it can for acorns and orchids. Our natures are too plastic and our potentialities too vast for that."[17]

Utilitarian considerations may be less problematic when invoked in the context of legislation as opposed to everyday life decisions. Graduated income tax is generally justified on utilitarian grounds, as are fees levied by homeowner associations for the upkeep of communal property such as parks. Use of eminent domain to appropriate a person's private property for building a much-needed school depends on weighing of pros and cons in a utilitarian spirit, as is the requirement that dogs be vaccinated against the rabies virus. More generally, legislative proposals are often assessed by calculation of who benefits, who loses, by how much, and the relative numbers of those gaining versus those losing. Lopsided cases are the easier ones—those where many benefit enormously and a few bear a small cost, or where many bear a minuscule cost and some reap a substantial benefit. Nevertheless, in some cases the utilitarian calculation clashes with other values, yielding morally repellent recommendations.

The cases that prove problematic for utilitarians are those where the math specifies which option will maximize aggregate utility but that option will trash dearly held values, such as the value of punishing only the guilty, the value of private property, or the value of an individual's life. Why not produce babies to harvest their organs, assuming only a few are sacrificed and many are saved? The utilitarian math bids us go ahead. Stopping an insurrection by pretending to have found the leader and then making a public spectacle of his execution can rate high marks on the utilitarian math. (My Russian friends call this *Lenin's math.*) Knowingly convicting an innocent person is morally repugnant, utilitarian math notwithstanding.

Tyranny of the majority against the wishes of the minority is a complaint sometimes leveled against utilitarians who are ready to restrict speech and freedom of the press during political turmoil or who assume they know what is best in the long run for all of us. As Simon Blackburn ruefully points out, "Just as a lot of crimes are committed in the name of liberty, so they can be committed in the name of common happiness."[18]

Utilitarianism and Kantianism, the two leading ethical theories, look a bit thrashed at this point, though none of the criticisms outlined here are new. What has gone wrong in the philosophical domain? I will hazard a hypothesis.

CONSTRAINT SATISFACTION AND MORALLY DECENT HUMANS

The ethical theories examined in the preceding discussion seek a clear, simple system that will not founder on messy exceptions and irresolvable disagreements. They aim to settle differences of opinion about what is right by referring to a rule or a set of rules whose authenticity they believe they have amply demonstrated by reason alone. The common flaw is that each approach selects as fundamental one kind of constraint in the decision-making process, such as the expected consequences for happiness of all (the utilitarians), or rules derived from reason (the Kantians). They then try to hoist their favored single constraint as the be-all and end-all in their "ethical theory of right action."[19]

It makes more sense to acknowledge that usually many constraints are relevant to a decision. You don't have to be a utilitarian to know full well that the consequences of a choice are always a relevant constraint. The assessment of the impact on oneself and on others is always important. Likewise, you don't have to be a Kantian to know that widely accepted rules are a relevant constraint. Sometimes, but not always, a familiar maxim may come to mind—"Look before you leap," for example. Jumping into deep water to help a person who has gone overboard is a bad idea if you cannot swim or you have a broken leg. You might just end up making a rescue far more complicated. If you want to join Doctors Without Borders but you are leaving behind four young children and a very ill spouse, you might consider the old saw "Charity begins at home."

What does decision-making in the brain actually look like? Primate and rodent studies of neuronal activity during decision-making reveal that the brain accumulates evidence over both short and long timescales. The circuitry is sensitive to the reliability of evidence from different modalities, and it draws on spatial knowledge and on past experience in making a choice in a similar circumstance.[20]

Background drives such as the need for food are strong constraints, especially when an animal is hungry. In some experiments, the animal has to decide how long to continue to forage in a patch as its resources are depleted, and when to move on to a patch whose value is uncertain. Values are compared in the rodent's brain, and the animals also estimate their own confidence in a decision.[21] Remarkably, it appears that the animal's decision is typically close to optimal, as determined mathematically. When suboptimal decisions are made, it turns out that, in both rats and humans, noise in the sensory systems or reliance on irrelevant information is often to blame.[22]

In human choice generally, behavioral and brain-imaging studies indicate that similar neural operations are involved.[23] For example, humans, like other mammals, recognize a relevant similarity between the case at hand and other cases encountered sometime earlier in life. Psychologists call this *case-based reasoning.* Since we use case-based reasoning for many problems in the physical world, it is highly probable that we use it in the social world, and this conclusion is confirmed by behavioral research. Like rodents, humans making a decision have a rough sense of how confident they are in the value of one option versus others. Like rodents, humans recognize that some evidence is more reliable than other evidence.

Moral decision-making also involves having a sense of how respected others in the community would regard the case, as well as assessing such matters as resources, and the capacity for follow-through.[24] In cases of serious moral issues, people tend to seek the opinions of others.[25] Many constraints jockey for consideration, and an introverted person may evaluate a constraint differently from an extrovert. In addition, different cultures may evaluate a particular constraint as more or less worthy. In the farming community of my childhood, for example, hard work and thrift were highly valued, while laziness and extravagance were disdained. We did realize, though, that other communities rated these traits rather differently, valuing instead artistic pursuits or athletic accomplishments. Even so, the adults were not especially judgmental concerning such differences, reckoning them as just the way things are. Innate constraints will reflect an individual's attachment to their offspring or mates, or perhaps their need for safety or food. Neurobiology cannot yet tell us exactly how neural networks implement procedures to satisfy constraints, but promising models do exist.[26]

Utilitarians recommend that we narrow our moral focus to only *one* constraint—maximizing aggregate happiness—but such narrowing is at odds with how the brain actually makes optimal, or near-optimal, decisions. Essentially always, a host of constraints are in the decision-making mix.[27] You may wonder whether faithful adherents of utilitarianism are actually gumming up their brains' highly evolved procedures for constraint satisfaction by giving undue weight to a single constraint.

Careful observations of sociality in nonhumans, both in the field and in captivity, regularly reveal actions that in humans might well be called "moral."[28] At the very least, these data suggest there are evolutionary precursors to human morality. This question is well worth pondering, because until the last few decades, a common assumption, at least in the Judeo-Christian West, was that only humans show moral behavior. The beasts are just beastly.

Consoling a friend after a defeat, cooperating to achieve a goal, sharing food, reconciling after a squabble, punishing an individual who harms others, adopting an orphan baby, and grieving after the loss of a loved one—all of these behaviors have been seen in chimpanzees and bonobos both in the wild and in captivity.[29] Many are seen also in baboons, wolves, monkeys, and rodents. Rats, I was surprised to learn, will show prosocial nonselfish behavior. In one study, rats that controlled access to a food cache would grant their hungry friends access to the food, forgoing the benefit of getting the food themselves.[30] Rats turn out to be more socially forthcoming than I had assumed.[31]

While consolation of the troubled is a prototypical feature of moral behavior, different situations call for other kinds of support. Conflict with an out-group requires not consolation, but cooperation and loyalty. Everyone needs to be able to count on the support of everyone else. Does oxytocin figure in this kind of support? Apparently so. A striking set of results comes from observations of wild chimpanzees.[32] When, as does happen from time to time, there are intergroup conflicts between chimpanzee troops, tight cohesion and coordination among members within the group, regardless of social affiliation, are observed. Their lives depend on it.

By collecting urine samples (in the wild) and measuring oxytocin in the urine, researchers found that oxytocin levels are higher for individuals of both sexes immediately before and during the conflict. It is a reasonable guess that oxytocin is important in enhancing feelings of bonding to group members. Because animal studies show that oxytocin also reduces anxiety responses, it is a question whether it does so even in conflict situations. Bravery, for example, might be boosted along with group cohesion. The results of some human studies where oxytocin is sprayed up the nose do indicate stronger within-group bonding, accompanied by greater hostility to those in the out-group. Intriguing as the results are, methodological worries recommend caution for now (as discussed in Chapter 2).

Following other contemporary moral philosophers, such as Mark Johnson and Owen Flanagan, I have come to view the prospect of a clear, simple rule or set of rules that holds for all persons in all situations as undermined by the reality of social life. While most humans may agree on a set of core examples of actions that are morally wrong, things become much more problematic at the fuzzy boundary (see again the Introduction). Moral judgments are not like mathematical judgments: 2 + 2 is always 4, but telling the truth is not always the morally best decision, even though it often is. Although there is no rule for when it is morally preferable not to tell the truth, in general, the human brain is surprisingly adept at computing when that is.

Aristotle and Confucius stressed the importance of developing strong social habits, also known as the *virtues*: prudence, compassion, patience, honesty, courage, kindness, hard work, and generosity. All habits reduce the costs of decision-making. As we have seen, brains aim to keep their energetic costs as low as is consistent with well-being, and habits are one good solution to energetic efficiency. The advantage in cultivating virtues such as compassion and honesty is that, as habits, they bias the constraint satisfaction process in the direction of morally decent decisions—decisions suitable for a highly social mammal such as *Homo sapiens*. Though such habits are never the be-all and end-all in moral decision-making, an ingrained virtue such as kindness means that the brain does not have to compute and evaluate from scratch all the factors relevant to the choice.

In other words, if you have a habit of being kind to everyone, you do not have to use time and energy figuring out what to do in a routine occasion. If something highly unusual were to occur, the habitual response might be put on hold. As Aristotle rightly observed, the virtues do not inflexibly determine what you will do. In general, however, habits, whether social or otherwise, reduce the costs of constraint satisfaction operations because fewer options are considered and hence fewer computations are required to find the optimal outcome for the situation at hand. When conditions are dramatically different from the norm, a bias in favor of kindness may be adjusted in favor of prudence and gathering further evidence, for example. By contrast, the computational costs of a systematic utilitarian's brain, making one decision at a time, may be excessive and wasteful, and you have to wonder how such a person ever gets anything done.

The additional Aristotelian point is that the moral environment is important as children grow up. It matters that decency, respect, and kindness are normal in the environment.[33] The opposite traits cause conflict, and a breakdown of the trust needed for cooperation in tough times. Developing socially considerate habits makes life easier and better for everyone. Habits do not dispense with judgment, but they do reduce the energetic costs of decision-making.

MORALITY FOR HUMANS[34]

A compelling merit of the biological approach is that it leads us to a plausible way of understanding why we can be motivated to behave morally at all, why acting kindly or generously does not go against our nature, and why virtuous habits are valuable. It helps us understand why even preverbal children are apt to show empathy and will spontaneously try to help.[35] It explains why we value a social life and the benefits it brings, even when there are some costs. It suggests an explanation for why approval and disapproval powerfully motivate individuals to learn the social practices of their group, and to get along in their group.

Can we get closer now to a precise definition of morality? No, for the reasons outlined in the Introduction. *Morality* shares with many of our everyday concepts a radial structure, meaning that it has uncontroversial cases at its center, and radiating out are cases with declining similarity to the core cases. The boundary is fuzzy. Nevertheless, we may find that a first-pass formulation is good enough to smooth discussion: *Morality is the set of shared attitudes and practices that regulate individual behavior to facilitate cohesion and well-being among individuals in the group.* Social practices regarding how to get along create expectations that figure into decisions about what to do. Expectations concerning how others will almost certainly behave and react entail energy-relevant efficiencies in decision-making, and failed expectations may generate a sense of something being wrong or at least amiss.

These expectations and feelings may be consciously apprehended, or they may not be. Even among verbally fluent humans, the social practices followed may be not articulable. Who among us can say exactly how close to stand to a new acquaintance, though our brains recognize when the distance is too great or when it is too small, and they give us an anxiety signal. Who can say exactly what behavior by a fellow passenger is too friendly, though most of us know quite well when the line of propriety has been crossed. Children usually pick up from observation that certain topics, such as finances, are not discussed outside the family, and certain other topics are not discussed even within the family. In the same way, they learn when and how laughter can be used to lower the temperature of an awkward situation and when to withdraw from a discussion that has become overheated.

Moral norms emerge in the context of social tension, and they are anchored by the biological substrate. Learning social practices relies on the brain's system of positive and negative reward, but also on the brain's capacity for problem solving. Just as hominins learned how to build boats suited to their particular ecology, they also acquired and honed social practices that helped the group prosper in its particular ecology. And just as each generation of boatbuilders may invent a slight modification to the standard design, social rules may be modified with the emergence of new conditions and new ideas. Not every idea to improve a boat actually yields a better boat. Likewise, not every idea to improve social norms actually results in a worthier outcome. Progress cannot be counted on, however much it is hoped for.

Are there universals of boatbuilding? Not really, but perhaps sort of. No clan builds boats out of rocks; no clan cuts holes in the bottoms of boats they intend to use. I surmise that no clan ever needs to spell this out for budding boatbuilders. Beyond the obvious, design is shaped by the kind of water the boat will be on—turbulent rivers, rough seas, calm seas, or glassy lakes. Many kinds of material can be used, from bark to logs to sealskin. In similar fashion, we can see undeclared universals in social norms. As the political scientist James Q. Wilson pointed out, moral norms that are universal do not need to be articulated as a law.[36] Beyond the obvious, the kind of moral norms a group settles on will be shaped by the social and ecological conditions, with the basic platform of strong desires for sociality always present as a background force.

For big-brained primates, such as the hunting-gathering-scavenging hominins, social practices reflecting preferences for stability, safety, and prosperity emerge in the course of daily life. Exactly how such practices develop is not well understood neurobiologically, though we have a good idea of the brain regions that play a prominent role. Just as norms often emerge in domestic arrangements without much explicit discussion, they probably also emerge in larger groups. Social practices perceived as successful are learned by youngsters and become part of the group culture. Later modifications would emerge similarly, perhaps without fanfare or much in the way of collective brow furrowing. Anthropological descriptions of the ways of small groups living off the land give us a feel for how early *Homo sapiens* likely managed their social lives.

Franz Boas's detailed descriptions of his early (1883–84) visits in the Arctic with the Inuit villages on Baffin Island illustrate the point that highly resourceful humans living in small groups have developed time-honored but modifiable social customs that foster stability and survival.[37] For example, food sharing by returning hunters happens seamlessly, as do customs about who does what work and in what order, how giving birth is to be managed, and what procedures to follow upon a death from old age versus a death from infection. Within-group fighting is strongly discouraged, and among the Inuit, song duels are often used to settle ordinary disputes.[38] Play is also a strong element of Inuit life, serving to strengthen social bonds, as well as to soften the sharp edges of annoyance and anger.

Among the Inuit of the nineteenth century, a single murder could be avenged by the victim's family, but then the cycles of vengeance had to stop. Someone who committed multiple murders required more attention. In one case that Boas describes, a man of one village took it upon himself to consult with those in other villages to determine whether the repeat offender should be executed. With wide agreement, the execution was carried out. The head man of the Akudmirmuit village went hunting with the offender, and shot him in the back. Other anthropologists report similar customs elsewhere in the Arctic. Nothing is written down, for the Inuit did not have writing at that time. Nonetheless, the understanding of how things should be done is picked up by everyone, stories and songs contributing to the viability of the custom.

These social arrangements by small hunting-fishing groups such as the Inuit suggest that basic moral customs are sensitive not only to the local ecology, but also to the needs and desires and hopes of the people in the group. Just as the Inuit invented remarkably tough and seaworthy kayaks that were essentially perfect for their needs, so their many social customs seem also to fit well with the basic mammalian-biology business of giving birth, hunting, keeping warm, and managing obnoxious behavior within the group, all the while living in the unforgiving Arctic. Just as they found ways of using sealskin in making leakproof kayaks, they also found ways of handling rare but necessary executions. Despite the harsh conditions, including starvation, long-lasting storms, and the death of dogs from contagions, the Inuit flourished. Those who see such lives as morally unsophisticated might want to take Boas's advice and look more closely.

Reports by anthropologists indicate that bands of hunting-gathering-scavenging *Homo sapiens* living 250,000 years ago were probably not unlike the Inuit visited by Boas. Their flexible customs were passed on through an oral tradition, and were largely suited to their ecology and way of life. The relative simplicity of their circumstances and their small band size (about thirty to fifty) meant that even though their social practices provided guidance, they were not hard-and-fast rules. Genetically, early *Homo sapiens* were essentially the same as we are now, save for some differences in genes for traits such as digestion enzymes and hair texture.

As far as is known, the first explicit codification of rules was completed by Ur-Nammu, King of Ur (2050 BCE) in Mesopotamia. The more famous code of Hammurabi in Babylonia is dated somewhere between Hammurabi's presumed dates of 1792–1750 BCE. By that time, domestication of animals and cultivation of crops had been under way in the Middle East for about eight thousand years. Large, stable communities had developed, and in those conditions, many aspects of social life had changed from Stone Age times. Kings, slaves, and private property became common institutions. Hammurabi's code contains rules to govern contractual agreements, punishment levels for types of offenses, inheritance, divorce, military service, who is responsible if a house falls down, and prices for goods and services (e.g., by ox drivers and surgeons).

Importantly, then, for most of the history of *Homo sapiens* on the planet, we were living in small bands, moving from place to place to follow game and adapt to climate challenges. Our ancestors relied on customs regarding very basic features of social life, much as we know the Inuit, the Comanche, and the Haida did until very recently. Dwelling on this fact prompts us to recognize the rather close fit between the human neurobiological platform that undergirds our caring for family and friends, and the set of customs that can arise on that base to cope with the many demands in the physical and social environment. And it is in this fit that we can make sense of moral norms—not as things apart from our nature, as things foisted on our nature, but as practical solutions to common problems.

When we look at our current laws governing complex institutions such as a congress or a criminal justice system, the fit between neurobiological platform and cultural norms may seem so indirect as to be nonexistent. As a result, we may slip into seeing such laws as reflecting "real" or at least "higher" morality, and as offering the best model for what morality really is. We may even convince ourselves that our advanced culture has found "stable moral truths" that are independent of our biology. More likely, however, it is the outcome of just more problem solving in the different ecological conditions of the industrial and postindustrial ages.

THE VOICE AND FEELINGS OF CONSCIENCE

Is there not a special feeling of "oughtness" that accompanies only moral choices, and not mere customary choices?[39] The voice of conscience, we might say. Just on the basis of that feeling, can we not tell directly that morality stands on wholly different ground from social norms? As we saw when discussing the reward system, feelings of "oughtness" (positive valence) or "ought-not-ness" (negative valence) typically accompany social habits that emerge from reinforcement learning, as well as from imitation, which also engages components of the reward system. Such feelings can be very strong and demanding, or weaker and niggling.

How strong these feelings are depends on what we have learned about how to behave, on the consequences of one action rather than another, and on the degree to which our emotions are engaged. It also depends on our judgment about how much social disapproval would ensue if others knew about the decision. But I am skeptical of the idea that a unique neural network is engaged precisely and only for the moral oughts.

Here is an anecdote: As a child, I had Catholic friends whose "ought not" feelings were as powerful when they contemplated eating meat on Friday as when they contemplated a genuinely moral issue such as fibbing or making a mean joke about a girl's buckteeth. Even when the dietary restriction requiring fish on Friday was dropped by the church, my friends continued to be awash in feelings of ought-not-ness and to dread being tempted by a Friday meatball. The reward system's ingrained habits do not change as quickly as a religious policy may.

Strong feelings may accompany adherence to or lapse from certain dress codes, such as not wearing a head scarf or not covering one's breasts. These codes are cultural; they are certainly not innate. Remember those *National Geographic* photos of women in villages who modestly covered their legs but were entirely comfortable baring their breasts? How amazed we all were as adolescents that a different custom could so modify feelings. It did not *feel* like a merely cultural requirement to cover one's breasts. Indeed, to us farm girls, exposure of breasts, especially in the company of males, may have felt much worse than telling a "white lie" (personal experience)—ostensibly a moral transgression.

On occasion, rudeness—not letting a driver of another car merge into the lane—can inspire as much outrage as breaking a promise. For many years, driving home drunk from the pub was regarded as tolerable boys-will-be-boys tomfoolery, innocent deaths and grotesque wrecks notwithstanding. A few short decades of campaigning by Mothers Against Drunk Driving, and many tragic deaths later, the tide has turned. Now we view drunk drivers as moral cretins.

In sum, the same fundamental brain mechanism is at work whether a plan is assigned a powerful negative valuation or only a mildly disapproving appraisal (see Chapter 4). Our culture strongly influences whether we regard a practice as so important and essential to membership in the group that we deem that practice to be on the moral end of the spectrum, or whether, like table manners, the practice is regarded as toward the social-convention end of the spectrum. For example, Americans generally tend to feel very strongly about showing respect for their flag, considering respect a moral duty, while Canadians typically are much less emotionally connected to their flag, and regard respect for their flag to be a rather pleasant social convention. Cultural practices, and even the language we share, affect how we characterize *what* we feel—moral outrage or nonmoral indignation at rudeness, for example. Which description we use, and perhaps what we consciously feel, is context dependent.[40]

Wiring that is genetically ready at birth supports the disposition to care, which in turn supports the motivation to acquire the social practices of the community—to find disapproval aversive, and approval rewarding. Cultural practices, whether about food sharing, lying, marriage, murder, or generosity, are learned. Conformity to social norms meets with approval, and approval positively reinforces our decision. Transgressions meet with disapproval, and such actions consequently receive negative valence. Through structural changes, the brain thus alters the probability that such an action will occur again in that condition.

Once learned, social norms become part of a person's ever-developing extended neural network, in cortex as well as in subcortical structures. Memories, language, and imagination, along with bonding to family, friends, and tribe, will modulate and sculpt that neural network. Certain cultural practices may seem absolute and universal. To the beholder, they seem to be right, period. The reward system in social mammals is apt to foster convictions like that. Although the feeling of moral certainty associated with the norms of one's own group may be adaptive, we also know of independent individuals who challenge the certainties and disrupt the uniformity of deep-seated convictions. Thus did slavery come to seem wrong, and the banning of whale hunting come to seem right. Times change. Nonetheless, we cannot be so certain that all changes end up being instances of moral progress.

THE JOY OF BEING BIOLOGICAL

The aim of this chapter has been to let the biological approach to human morality enter the ring with more mainstream philosophical approaches. By *mainstream* I mean belief in a lawgiving God and pure reason, either together or separately, as the wellspring of moral understanding. Although I put my bets on biology—instincts, learning, problem solving, and constraint satisfaction—it is important to lay out the arguments for reason and religion, and to consider their merits, as well as their flaws.

The brain's reward system is known to play a powerful role in what we think we ought to do, morally and otherwise. That voice of conscience that we hear when we consider violating a norm is our reward system sending out a "negative value" signal. Our conviction that we are justified in a choice does not come from some hypothetical "pure reason" unconnected to the physical brain. It depends on what our brains have internalized as an appropriate norm—on what our reward system assigns value to, and which constraints dominate. Inner conflict may arise when our internalized values conflict—such as loyalty versus truth, or truth versus damage caused. And sometimes talk, either internal or with others, does not resolve the norm conflict. Often after a long walk or a long sleep, our conflict subsides and a decision is made. This is not just a neocortical function, let alone a function of "pure reason." It is a whole-brain function. The reward system and its assignment of valuation are crucial.

For most of my philosophical education, the unchallenged and prevailing view was that reward learning could not conceivably get us anywhere close to how, in fact, moral norms are apprehended and followed. Reward learning was considered hopelessly unsuitable to the nature of moral knowledge. Conditioning, we were told, is just dinky association of *this* with *that*. Knowledge of moral norms goes way, waaaaaay beyond that.

The problem with this bit of philosophical certainty is that it was wrong. Dead wrong. It was based on ignorance of how subtle and clever and powerful the reinforcement learning system is in mammals, especially as interconnected with the cortex and the hippocampus. The thing about the human brain that is special is its very large number of neurons—86 billion. A macaque monkey has about 6.5 billion neurons.[41] How our 86 billion neurons work together so that we talk and do fancy math, so that we create symphonies and parliaments, is not known. Nevertheless, having many more neurons appears to be one factor that makes the human brain a little different from other primate brains.

It is a good guess that when more neurons are added, behavioral capacities scale up—not linearly, but *exponentially.*[42] This is because any given neuron will make about 10,000 connections with other neurons. Ten times the number of neurons means, roughly, a 10^{10} increase in fanciness, very roughly. The remarkable fact is that there is no secret structure unique to the human brain that creates sophisticated social institutions or art or a moral conscience. Our brain just has more neurons.

In all mammalian species, genes build the wiring and manufacture the suite of neurochemicals to ensure that the social life of individuals naturally includes caring for those to whom we are bonded.[43] Yes, human sociality is different from that of chimpanzees, whose sociality is different from that of bonobos, or that of prairie voles. Nevertheless, the basic principles, involving both what I call the platform for morality and the powerful capacity of the reward system to support acquisition of cultural norms, appear to be broadly shared among highly social mammals.

One of the joys of biological evolution is the variability across species, but also within a species. For highly social animals, the inescapable variability means that within a group, there are differences in temperament, in values and preferences, and in styles of getting on in life. Inevitably, individuals compete, and just as inevitably, they rub each other the wrong way. After some noisy displays, things generally settle down again, and if good sense prevails, significant disagreements can be negotiated to resolution. Variability among individuals also means that there is novelty and creativity, with the result that there is much to be enjoyed, as well as much to be learned, from how others solved a problem, a problem about growing potatoes or herding sheep or a how to ensure that a contract for future goods is kept. Biological variability also means that in interactions between groups, social wisdom may be far more effective than trying to apply one's favored moral rule and then being morally outraged when the outcome is not what was hoped for.

Among Romans, the great emperor Marcus Aurelius (161–180 CE) was a wise ruler and a sensible person. The last word belongs to him:

> Live a good life. If there are gods and they are just, then they will not care how devout you have been, but will welcome you based on the virtues you have lived by. If there are gods, but unjust, then you should not want to worship them. If there are no gods, then you will be gone, but will have lived a noble life that will live on in the memories of your loved ones.[44]

CHAPTER 8

The Practical Side

The most practical kind of politics is the politics of decency.

THEODORE ROOSEVELT

Up to this point, the story of conscience has privileged its social slant—when care and concern are directed toward others. Focusing on the social slant makes sense, since that is where our conscience may urge us to act against our own interests for the sake of others. Historically as well, philosophers and theologians predominantly considered the role of the conscience in how we get along with others, sometimes with others that are alleged to be gods.[1] Not unusually, however, we may think of conscience as also weighing in on duties to ourselves. Conscience may have a voice when it is mainly we ourselves for whom the consequences matter.

Moderation in eating and drinking, for example, or the development of traits such as patience and prudence, are among the virtues that our conscience bids us aim for. When we falter in our resolve to exercise regularly, for example, our conscience is apt to provide a rebuke. Such self-oriented duties as moderation in food can also have some indirect effects on others. Ill effects on health that result from overimbibing, for example, often adversely impact family and friends. Nevertheless, the focus here is primarily our own well-being or achieving our potential.

Socrates, though famously oriented toward social issues, was nevertheless unwavering in this apparently simple piece of advice: *Know thyself.* He was not urging us to be self-absorbed. Rather, Socrates was warning against the many ways in which we may fool ourselves, even when the unconcealed facts are staring right at us. Socrates knew that when we are not firmly honest with ourselves or when we indulge in wishful thinking, we make ourselves vulnerable to exploitation. When we do not admit our mistakes, we risk disaster. Living, as he was, during a time of catastrophic wars and political upheaval, Socrates was painfully aware of the rot that spreads when citizens willingly allow themselves to be bamboozled by the powerful.

What really got Socrates's goat was arrogance. He noticed that especially in the moral sphere, individuals often pose as knowing much more than they do—as having profound understanding when, in fact, their knowledge is measly. Pretenders to wisdom were a favorite target for Socratic inquiries. Steadily but gently asking questions, Socrates deflated the pomposity he abhorred. What he thought we should appreciate about ourselves was how vast our ignorance is and how easy it is to blunder by overconfidence.[2] We cannot learn anything if we are satisfied in the illusion that our knowledge is complete, or nearly so. We must have a realistic assessment of our strengths and our weaknesses to skirt pitfalls and seize opportunities.[3]

Certitude, Socrates thought, is often a dangerous affectation, especially in politics. We do well to be skeptical when we see certitude on display. In his view, we are easy chumps unless we are on our guard against certainty supported by rousing rhetoric but unsupported by genuine knowledge. He saw this vulnerability as vexing not only the political domain—for Socrates the high stakes were the future of Athenian democracy—but also religion, philosophy, and the practice of the law. Socrates was having none of it.

Know thyself. Socrates seems to have believed that this simple point expresses our most fundamental duty to ourselves. His advice can seem challenging when the comforts of myths and magic mesmerize us. The truth is sometimes brutal. Regardless of how much we conceal the truth, or pretend it is not so, the data are the data.

Wait. Socrates's exhortation is all beside the point unless we really can make choices. Unless we really can inhibit unwanted impulses, unless we really can choose to decline the soothing myth and embrace the harsh reality, Socrates's admonitions are just so much hot air.[4]

Some scientists and philosophers do, indeed, conclude that controlled choice or, as you might say, free will, is itself a myth—and a myth we ought to dispense with.[5] The argument goes essentially like this:

> *Free will requires acting in a causal vacuum. In other words, a truly free choice must not be a caused choice. The brain is a causal machine, designed by the genes, and all our actions are the outcome of brain operations. All our choices and decisions are caused by processes in the brain. Therefore, there is no such thing as free will.*

A corollary might be this:

> *Because no one has free will, no one is truly responsible for anything. Therefore, the criminal justice system should be dismantled.*

This is a momentous conclusion with very serious social consequences. If a psychopath is also a serial rapist and murderer, is the recommendation that we just sigh and say, "Well, it is not his fault he is a psychopath. There is no justification for locking him away"? Or do we lock him away and simply say that he is not responsible, but merely abominable? Or what?

Three points are relevant here. The first concerns the rationale for having a criminal justice system at all. The second concerns whether treatments are available for turning psychopaths and hardened criminals into socially decent human beings. The third addresses the semantic point that free will *means* choosing in a causal vacuum. I will take these in order.

Many thinkers, from ancient times until now, have recognized the practical basis for a criminal justice system.[6] One fundamental point is pragmatic: the criminal justice system is anchored by the need for social safety and security. Some individuals can be extremely dangerous, such as someone who repeatedly murders other humans for the sheer fun of it. A second fundamental point is that rough justice will prevail unless a reasonable system exists for correctly identifying offenders and then coming to fair-minded decisions about what to do with them. When no judicial system prevails, the family and friends of the victims, for example, will probably take the law into their own hands. Rough justice is typically very ugly indeed. Any legal system will have flaws and imperfections, but typically they pale in comparison to what vigilante justice delivers, meting out death and destruction as easily to those who are completely innocent as to those who are actually guilty.

This practical point implies that the moral high ground cannot be claimed by those who advocate dismantling the criminal justice system because the brain is a causal machine. Disappointingly, those who argue that free will is an illusion have little in the way of substantive suggestions for improvements to the system, especially improvements consistent with the pragmatic considerations emphasized here.

"Treatment is surely the better alternative to punishment" is a common response to the pragmatic rationale for a criminal justice system. Worthy though this suggestion surely is, the problem is again a practical one: Are treatments available? What are they? The fact is that no effective treatment for psychopathy exists. And the same is true for the criminally antisocial. So how do we prevent a repeat offense by a psychopath who has already killed multiple times if he cannot be removed from society? Recall that about

25% of prison inmates are diagnosed as psychopaths. Many more of the prison population have antisocial personality disorder.

Precisely because prison is punishment and is undesirable for the prisoner, our good natures find it appealing to conceive of a magic pill that would just transform a psychopath into an empathic, decent, law-abiding citizen. This idea is, alas, worse than far-fetched. Science is nowhere close to having such a pill or even meaningfully experimenting with one. Moreover, an educated guess suggests that discovering such a pill even in the long run is improbable. By all means, let's aim for that treatment, just in case we get lucky. In the meanwhile, let's not pretend that the treatment is in hand, or even close.

On to semantics. What does the expression *free will* mean?[7] The semantic claim that having free will *means* that our choices and decisions must be made in a causal vacuum is inaccurate. As the philosopher Eddy Nahmias has shown experimentally, ordinary speakers of the language do not mean by *free will* that you choose in a causal vacuum. They mean, among other things, that you have knowledge, intent, and self-control.[8] Of course, meanings of words do shift and change, and perhaps the meaning of *free will* should change to reflect the idea of a causal vacuum. If so, a convincing argument for a meaning change is needed, but first it must be reconciled with David Hume's insights.

Hume argued that it is not causality per se that excuses one from responsibility. Rather it is certain *kinds* of causes. For example, suppose you are thrown from a horse and as a result knock over a lantern, setting the barn on fire. We say the fire started accidentally, not intentionally. Or suppose you are so delusional that you club the bus driver because you believe he is actually Adolf Hitler masquerading as a bus driver. These are cases in which certain kinds of causes are absent. Absent are some of the following: the intention to perform the action, self-control, reasonably accurate knowledge of conditions, and conscious awareness of what one is doing.

As Hume would insist, the important question is what *kinds* of causes diminish self-control or impede knowledge, and whether the criminal justice system takes such differences in causal antecedents into account in judging responsibility. To which the fast answer is yes, it does. In his wisdom on these matters, Hume was preceded by Aristotle, who made many of the same points. As Aristotle commented, there is a major difference in responsibility between someone who drinks too much and reveals state secrets, and someone who is involuntarily intoxicated and reveals state secrets. It would be futile to punish the latter. Similarly, someone who is completely delusional may need to be in custody, but punishment is futile.

Now consider Bernie Madoff.[9] He was a financier in New York who ran a very slick Ponzi scheme for some twenty years and bilked people out of about 65 billion (with a *b*) dollars. Following the economic crash of 2008, many of Madoff's investors wanted to take their money out of his fund, but the fund, being a Ponzi scheme, collapsed. Confronted by his sons, Madoff finally confessed his crime, and they notified the police. Here is the thing. You cannot really run the biggest Ponzi scheme in history without knowing what you are doing, exercising extraordinary self-control, and intending to do what you are doing to keep pulling in new investors and paying out 10% per year to existing investors.

Madoff exercised self-control and planning and cunning to avoid being caught. He was not criminally insane. He was not coerced. No gun was pointed at his head, his family was not starving, there was no extortion. Like everyone else, Madoff has a brain that is a causal device. But so what? No jury would ever be persuaded that Madoff ran his Ponzi scheme by accident or that he did not know what he was doing. No mention of free will needs to be made here. To hold Madoff responsible, it is enough that he had self-control and intent and knowledge, and that he had no legitimate excuse. That is what common sense requires and also what the law requires.[10]

Does "free will is an illusion" mean that there is no difference between a brain with self-control and one with no self-control or one with diminished self-control? To assume there is no difference related to self-control is flatly at odds with the neurobiological facts. Self-control is a brain function that develops during childhood and is susceptible to deterioration from injury and drugs and conditions such as addiction or meningitis. Rodents show self-control in the sense that they can defer immediate gratification for a larger later reward; they can stop an action once started, they can inhibit self-defeating desires, and so on. And I *am* talking about rats here. It strains credulity to maintain that no human ever has self-control. On the contrary, humans appear to have an even greater capacity for self-control than rats have. We can defer gratification for very long time periods, stop an action once started, and ignore distractions while we pursue a goal. Not perfectly, not always, but for most humans, regularly enough.

Basically, if you intend your action, know what you are doing, are of sound mind, and are not coerced into the decision (no gun is pointed at your head), then you are responsible for your action. These simple requirements of responsibility are generally good enough for practical purposes.[11] Importantly, legal opinions in criminal cases are often sensitive to the complexity of human life. If you killed in self-defense, for example, you can claim a mitigating circumstance. The sensitivity of judicial opinions in case law pertains to all three phases of a criminal proceeding: determining competency, determining guilt, and sentencing. Which of those three do the no-free-willers want to dismantle? All three? Only competency? Only sentencing? Only determination of guilt?

Sometimes the no-free-willers declare that they are okay with removing offenders from society and they want only that we should stop telling the Bernie Madoffs of the world that they are bad people and that they are responsible for the havoc and injury they knowingly caused. While such sentiment may be motivated by kindness and compassion, I occasionally detect a faint whiff of self-righteousness. If you wish to tell the Bernie Madoff that because his brain is a causal machine, he is not responsible and not a bad person, by all means tell him so. If that will make you feel better, then perhaps some good comes of it. Madoff's victims, however, may find the whole self-comforting exercise a bit too precious to stomach.

Finally, it is essential to hear from a legal scholar on this matter. I turn to Stephen Morse of the University of Pennsylvania Law School:

> Contrary to what many people believe and what judges and others sometimes say, free will is not a legal criterion that is part of any doctrine and it is not even foundational for criminal responsibility. Criminal law doctrines are fully consistent with the truth of determinism or universal causation that allegedly undermines the foundations of responsibility. Even if determinism is true, some people act, and some people do not. Some people form prohibited mental states and some do not. Some people are legally insane or act under duress when they commit crimes, but most defendants are not legally insane or acting under duress.[12]

Where does all this leave us with respect to our conscience and how it works? From a neurobiological perspective, it is tolerably clear that judgments of conscience involve a whole flock of interactions and integrations, valuations and attentional allocations, simulations and emulations, emotions and self-control. If we seek a precise explanation of why a seemingly decent person acted against conscience by taking a bribe or molesting a child or lying under oath, we have to acknowledge that neuroscience cannot yet provide such a precise neurobiological answer. Psychology cannot do much better. For one thing, each individual's wiring is unique, even though common themes and pathways exist. At this stage, to determine your particular wiring we would have to take your brain apart.

Within a decade or so, however, new techniques and new knowledge may enable us to say much more than we can now, at least in general terms. Even so, the neurobiological answer that will be available then is unlikely to be precise and complete, but only fairly general. With luck, it may be richer and more satisfying than merely saying that the miscreant is a bad person or that he succumbed to the devil or that his "id" overcame his "superego" or that he was temporarily insane.

Questions rooted in our gathering understanding of the neurobiology of our social lives fan out in many directions. Possible implications for criminal law, probably not in the immediate future but in the long run, engender one set of disquieting questions.[13] Another set of questions concerns clinical practice for those suffering psychological trauma that isolates them from the very social relationships that they need to heal.[14] Because information technology and social media are changing our social world, other questions arise concerning whether unhealthy forms of isolation are becoming more normal. What are the mental costs of social media if increased social isolation is the ubiquitous outcome?[15]

In addition, the results from political scientists and neuroscientists indicating that certain types of neurobiological dispositions, perhaps genetically linked, incline us to line up one way or another ideologically and politically raise a whole set of different issues. Could such knowledge make us more understanding of each other, or will it divide us yet further? Could it make us more resolved to craft practical solutions to disagreements, or will it mean that people will find it easier to write each other off as moral fools?

One enduring question is this: Who among us is a genuine moral authority? Who can we turn to when we want trustworthy resolution of a disagreement? The yearning for a moral authority is altogether understandable, since judgments from such an authority make our own decisions easier and our own conscience calmer. Socrates again intervenes, warning us of the legions of moral pretenders, especially those who claim they are above or beyond criticism. Socrates reminds us that we cannot escape responsibility by handing over control to a supposed moral authority. Socrates rightly feared despots who would galvanize us to do such things as burn witches or hang aristocrats or grab clubs and torches as we blame our troubles on the designated foe. Certainty is the enemy of knowledge.

Moral wisdom may be assumed to be found in academia, especially among the moral philosophers. Would that it were so, but my own experience belies that assumption. In the sheltered life of academic philosophy, practical wisdom may be in short supply, replaced either by endless dithering or unwavering adherence to a favored ideology. On the other hand, it should be noted that the Center for Practical Wisdom at the University of Chicago is, as its name suggests, dedicated to understanding what is needed for wise decision-making in the many aspects of living a complicated life.[16] The center's research output seems to avoid both dithering and ideological perseverance.

In any case, we are all acquainted with people who lack formal education but have reliably fair-minded and farsighted judgment, and largely avoid moral arrogance and corruption. There was much wisdom in Inuit camps, for example, even when there was no formal education at all. In all cultures and all societies, there tend to be some individuals who seem a little more grounded and a little more wise than others. But even those individuals are not wise all the time or in all circumstances. At most, they are just somewhat more wise than the rest of us some of the time. Woolly as it is, the best advice is perhaps to allow yourself broad life experience and exposure to the human condition in all its beauty and horror. Do your best, but even then you will make mistakes.

In some manner, we aim to find balance between trusting the judgments of others and withholding complete trust; between respecting community standards while acknowledging the flaws in those very standards; between trusting the deliverances of our own conscience, and recognizing that despite the best of motives, we will sometimes err and our conscience will churn. There is no algorithm, there is no rule, for finding that balance. But we can strive to continue learning from our social experience, and often that striving yields results. We do not precisely know what the brain does as it learns to balance in a headstand, for example. But over time, we get the hang of it. To an even greater degree, we do not know what the brain does as it learns to find balance in a socially complicated world.

Serious moral quandaries arise in wartime, at times of political turmoil, and during the aftermath of natural disasters such as earthquakes or hurricanes. My German-born friend Trudy was a toddler when Nazi Germany invaded Poland. Two years later, her father perished in the Battle of Stalingrad. Before he left for the front, he gave her mother a pistol, explaining that it was certain that Germany would eventually lose the war against the Russians. When the Russian army advances to Berlin, he predicted, they will naturally be seeking revenge. To avoid a terrible fate, he told his wife, she must shoot their two daughters and then herself. Unthinkably painful as this scene is, Trudy's father's directive was prompted by tremendous love and overwhelming fear, as well as despair over the insanity of the war. As we know, he was right about the outcome of the war in the east, but with cunning and determination, Trudy's mother managed to escape with her daughters to the west of Germany after the fall of Stalingrad, dodging the advancing Russian troops by hiding in tunnels and a salt mine, and going for days without food. Trudy recalls the unalloyed thrill of finding an old turnip. Abandoning the two children was never an option.

Reports from Japan after the March 2011 tsunami described shoppers quietly putting groceries back on the shelves and buying as little as they needed so that there would be food for others. People waited patiently in long queues to use the one public phone available, and others opened their homes to those who had nowhere to go. The spirit of community was widespread and heartfelt, and was deeply rooted in traditional values. It not only aided others in basic material ways, but reportedly augmented bonding between people, itself a positive feature that enhances resilience and motivates greater cooperation.[17]

Though catastrophes are not unusual on a worldwide scale, in most of daily life the many small acts of kindness, generosity, and courage add up to make a meaningful difference in the lives of others. Maintaining the virtues of decency and honesty is also extremely important in how people handle setbacks and tragedies, and how they respond to others in troubled times. Much has been written about the recent trend displayed by US celebrities and politicians in their disdain for social virtues such as honesty, kindness, and decency. These individuals are widely admired for such decency-scoffing behavior, so the flaunting of virtues spreads from a few to many.[18]

There is a social cost when those in influential positions venerate selfishness and boorishness. Public trust, an important value in a healthy society, is undermined when corruption becomes normal at the national level. Trust is what allows us to negotiate with one another despite our deep differences, and when trust evaporates, all we are left with are our differences. As we saw from the study on norm changes in the Montague lab (see Chapter 4), these changes to decency and honesty may occur without our awareness.

None of us are morally perfect, but the level of dishonesty and the lack of shame among many of our leaders predicts disorder and heightened conflict. Norms are being shifted away from virtue and toward greed and hard-heartedness. There are many exceptions to this trend, at all age levels, and the courage of such people is admirable.[19]

I AM WATCHING the summer dawn break over the mountains across the Queen Charlotte Strait in British Columbia. The news reports continue to show Honduran families who seek asylum in the US being detained at the border with Mexico. Parents are separated from each other; children are taken from their parents and put in camps that resemble cages. Babies and toddlers are pulled from their mothers and placed in "tender care" units. The policy behind these actions is called "zero tolerance" of illegal border crossing. No one knows how long the detainees will be kept in the facility, or whether any procedure is in place to reunite mothers with their babies. Some teenage children have already been detained for a month, and parents fear they will be deported without their children. Reporters are barred from entering the children's detention facility, so far. The attorney general has acknowledged that inflicting misery of this kind is a necessary feature of the policy to deter asylum seekers from trying to enter the US.[20]

I feel that recognizable tightening in my viscera. I must do something, but the options are limited. Once again, my conscience is badgering me.

Ideology, and the boundless enthusiasm for explaining away all moral reservations in the name of ideology, is what I most fear in the social domain. Certainty in the correctness of one's moral stance is what deeply troubled Socrates, and it troubles me now. My thoughts turn to Aleksandr Solzhenitsyn (1918–2008), the brilliant historian and novelist, who, for criticizing Stalin, was consigned for eleven years to forced labor in the camps of the Soviet Gulag.

> Ideology—that is what gives evildoing its long-sought justification and gives the evildoer the necessary steadfastness and determination. That is the social theory which helps to make his acts seem good instead of bad in his own and others' eyes, so that he won't hear reproaches and curses but will receive praise and honors. That was how the agents of the Inquisition fortified their wills: by invoking Christianity; the conquerors of foreign lands, by extolling the grandeur of their Motherland; the colonizers, by civilization; the Nazis, by race; and the Jacobins (early and late), by equality, brotherhood, and the happiness of future generations. . . . Without evildoers there would have been no Archipelago.[21]

Acknowledgments

Many friends lent a hand in writing this book. Roger Bingham and I have talked over the last two decades about topics ranging from morality to ideology to molecular biology, and I am grateful for his wisdom and breadth of knowledge. Roger was the first one to alert me to the importance of bioenergetic constraints in the evolution of the nervous system and in many aspects of cognition. He was way ahead of us all on that topic.

Deborah Serra and I regularly discussed over tea the role of the brain in moral decision-making and how individual differences in temperament, experience, and life situation make it inappropriate to have a one-rule-fits-all moral system. I am forever grateful for her sharp, critical mind, her piquant sense of humor, and her great kindness in reading rather clumsy early versions of this book and forthrightly recommending changes. Captain Dallas Boggs and Sue Fellows also kindly read the manuscript and tendered advice over many wonderful dinners.

As always, Paul Churchland was my in-house critic, copy editor, cook, pillar of strength, supplier of comic relief, and wellspring of common sense. He kept me going when I thought it might be better to forget the whole thing and work instead on expanding the raspberry patch.

Special thanks also to Read Montague and Larry Young, both of whom attended my 2017 Neuroscience School of Advanced Studies summer course in Tuscany on the origins of moral cognition, and generously discussed data from their labs that are directly relevant to moral decision-making and to the idea of a conscience. Dill Ayers, Ken Kishida, John Kubie, and John Hibbing all gave valuable feedback on earlier drafts, helping me to avoid really dumb mistakes. Mistakes may still exist, however.

Anne Churchland taught me about the neurobiology of decision-making, and how brains integrate a range of constraints at multiple timescales to make a decision. Her perspective gave me a new insight into why moral philosophers typically misunderstand decision-making in a moral context, and to appreciate more accurately the flaws in utilitarianism and Kantianism. In addition, Anne shared her hunches about what drove the early evolution of the cortex.

Mark Churchland has pondered the problem of the origin of moral motivation since childhood, and we have talked for many decades about these questions and about the Socratic caution regarding moral certainty, moral arrogance, and ideology. Mark helped me see the potential in a neurobiological perspective on social habits and skills, and how behavior and choice are tightly linked in the basal ganglia.

The senior editor at Norton, Amy Cherry, was helpful in a thousand ways, especially in providing advice on an early version of the book. She encouraged me to crank up clarity and smoothness, and despite my reflexive stubbornness, I am relieved that I acquiesced to her wisdom. A debt is also owed to the exceptionally efficient and knowledgeable copy editor, Stephanie Hiebert, who was a joy to work with. Thanks are owed to Woo-Young Ahn, John Hibbing, and Ken Catania for allowing me to use their illustrations, and to Julia Kuhl for drawing and redrawing figures for me. Finally, I wish to thank the Kavli Foundation for financial support.

Notes

Introduction: Wired to Care

1. I have changed the name of the village.
2. Royal Canadian Mounted Police, or Mounties for short.
3. Prime Minister Justin Trudeau officially apologized to First Nations people in 2017 for abuse. See "Justin Trudeau Offers Apology on Behalf of Canada for N.L. Residential Schools," YouTube, November 24, 2017, https://www.youtube.com/watch?v=p8CCAJzaT3I.
4. For some images of his paintings, see "Beardy, First Nations Painter," Google Images, accessed June 17, 2018, https://www.google.com/search?q=Beardy,+First+Nations+painter&safe=active&tbm=isch&tbo=u&source=univ&sa=X&ved=0ahUKEwiTyYKW_tDbAhWP7Z8KHfIKDjEQsAQIWA&biw=1002&bih=693.
5. Paul Strohm, *Conscience: A Very Short Introduction* (Oxford: Oxford University Press, 2011). This brief book is a splendid read and provides an excellent historical survey.
6. Strohm, *Conscience*, 2.
7. "Hearings Regarding Communist Infiltration of the Hollywood Motion-Picture Industry, House Committee on Un-American Activities," 82d Congress, May 21, 1952, in Ellen Schrecker, *The Age of McCarthyism: A Brief History with Documents* (Boston: Bedford Books of St. Martin's Press, 1994), 201–2.
8. Sheri Fink, "The Deadly Choices at Memorial," *New York Times Magazine*, August 30, 2009, 28–46.
9. Michael Stoltzfus, "Martin Luther: A Pure Doctrine of Faith," *Journal of Lutheran Ethics* 3, no. 1 (January 2003), https://www.elca.org/JLE/Articles/898#ENDNOTES.
10. "Letter to Frederick William, Prince of Prussia (28 November 1770)," in *Voltaire*

in His Letters: Being a Selection from His Correspondence, trans. S. G. Tallentyre (New York: Putnam, 1919), 232.

11. "Nation's Cancer Centers Endorse HPV Vaccination," Cold Spring Harbor Laboratory, June 8, 2018, https://www.cshl.edu/nations-cancer-centers-endorse-hpv-vaccination.
12. See Robert Wright, "Sam Harris and the Myth of Perfectly Rational Thought," *Wired*, May 17, 2018, https://www.wired.com/story/sam-harris-and-the-myth-of-perfectly-rational-thought.
13. Plato, *Apology*, in *Plato: Complete Works*, ed. John M. Cooper (Indianapolis, IN: Hackett, 1997).
14. Kwame Anthony Appiah, *The Honor Code: How Moral Revolutions Happen* (New York: Norton, 2010).
15. Ángel Gómez et al., "The Devoted Actor's Will to Fight and the Spiritual Dimension of Human Conflict," *Nature Human Behavior* 1 (2017): 673–79.
16. David Livingstone Smith, *Less Than Human: Why We Demean, Enslave, and Exterminate Others* (New York: St. Martin's Press, 2011).
17. George Lakoff, *Women, Fire, and Dangerous Things* (New York: Basic Books, 1987); R. L. Solso and D. W. Massaro, eds., *Science of the Mind: 2001 and Beyond* (New York: Oxford University Press, 1995).

Chapter 1: The Snuggle for Survival

1. Martin Nowak, an evolutionary biologist and mathematician at Harvard, coined this expression, to the best of my knowledge. For his contributions to understanding mammalian cooperation, see Martin A. Nowak, *SuperCooperators: Altruism, Evolution, and Why We Need Each Other to Succeed* (New York: Free Press, 2012).
2. The memorable expression "tend and befriend" was coined by Shelley Taylor and colleagues. See Shelley E. Taylor et al., "Biobehavioral Responses to Stress in Females: Tend-and-Befriend, Not Fight-or-Flight," *Psychological Review* 107, no. 3 (2000): 411–29, https://doi.org/10.1037/0033-295X.107.3.411.
3. "Adorable Friendship of a Shepherd Dog & an Owl," YouTube, September 30, 2015, https://www.youtube.com/watch?v=weL3N3W8VPg.
4. Sarah Blaffer Hrdy, *Mother Nature* (New York: Ballantine, 1999).
5. N. I. Eisenberger, "The Pain of Social Disconnection: Examining the Shared Neural Underpinnings of Physical and Social Pain," *Nature Reviews Neuroscience* 13 (2012): 421–34.
6. A. W. Crompton, C. R. Taylor, and J. A. Jagger, "Evolution of Homeothermy in Mammals," *Nature* 272, no. 5651 (1978): 333–36; Nick Lane, *Life Ascending: The Ten Great Inventions of Evolution* (New York: Norton, 2009). See also *The Origin of Minds* (New York: Harmony Books, 2002), by Peggy La Cerra and Roger Bingham, who were among the first biologists to really appreciate the widespread implications of energetic constraints.
7. Incidentally, this neurobiological feature of learning makes it difficult to imagine how a disembodied soul could learn or remember anything. What would encode the memories?
8. On birds, see Harvey J. Karten, "Neocortical Evolution: Neuronal Circuits Arise Independently of Lamination," *Current Biology* 23 (2013): 12–15. For the moment, I will pretty much leave birds out of the story. This omission is regrettable, but necessary, given space limitations.

9. Zoltán Molnár et al., "Evolution and Development of the Mammalian Cerebral Cortex," *Brain, Behavior, and Evolution* 83 (2014): 126–39; Jennifer Dugas-Ford and Clifton W. Ragsdale, "Levels of Homology and the Problem of the Neocortex," *Annual Review of Neuroscience* 38 (2015): 351–68.
10. MRI makes these neural structures visible. "Patricia Churchland's Brain," accessed July 10, 2018, patriciachurchland.com/gallery (scroll down).
11. Kent C. Berridge and Morton L. Kingelbach, "Affective Neuroscience of Pleasure: Reward in Humans and Animals," *Psychopharmacology* 199 (2008): 457–80.
12. Nathan J. Emery, "Cognitive Ornithology: The Evolution of Avian Intelligence," *Philosophical Transactions of the Royal Society of London. Series B, Biological Sciences*, 361 (2006): 23–43. The ethologist Bernd Heinrich is a careful observer of ravens: National Geographic, "Genius Bird," YouTube, July 11, 2008, https://www.youtube.com/watch?v=F8L4KNrPEs0. See also the TED talk by ethologist John Marzluff: "Crows, Smarter Than You Think" (TEDx Talks), YouTube, January 22, 2014, https://www.youtube.com/watch?v=0fiAoqwsc9g.
13. I should really say, *such structure as is visible to us through microscopes looks different in birds*. Not laminar. Nevertheless, the underlying—invisible—*principles* might well be similar. Genetic data and identification of cell types are beginning to suggest such similarity. Dugas-Ford and Ragsdale, "Levels of Homology."
14. K. D. Harris and G. M. Shepherd, "The Neocortical Circuit: Themes and Variations," *Nature Neuroscience* 18 (2015): 170–81; Peng Gao et al., "Lineage-Dependent Circuit Assembly in the Neocortex," *Development* 140 (2013): 2645–55.
15. Harris and Shepherd, "Neocortical Circuit."
16. As the anatomist Suzana Herculano-Houzel points out, if a mouse brain had as many neurons as a human brain has (about 86 billion), but with the typical packing density and neuron size of a mouse brain, that mouse brain would weigh about 36 kilos. That size is not workable. Suzana Herculano-Houzel, "The Human Brain in Numbers: A Linearly Scaled-Up Primate Brain," *Frontiers in Human Neuroscience* 3 (2009): 31, https://doi.org/10.3389/neuro.09.031.2009.
17. B. L. Finlay and P. Brodsky, "Cortical Evolution as the Expression of a Program for Disproportionate Growth and the Proliferation of Areas," in *Evolution of Nervous Systems*, 2nd ed., ed. Jon H. Kaas, vol. 3, *The Nervous System of Non-human Primates*, ed. Leah Krubitzer (Amsterdam: Academic Press, 2017), 73–96; Jon H. Kaas, "The Evolution of Brains from Early Mammals to Humans," *Wiley Interdisciplinary Reviews. Cognitive Science* 4, no. 1 (2013): 33–45.
18. L. Hinckley et al., "Hand Use and the Evolution of Posterior Parietal Cortex in Primates," in *Evolution of Nervous Systems*, 2nd ed., ed. John H. Kaas, vol. 3, *The Nervous System of Non-human Primates*, ed. Leah Krubitzer (Amsterdam: Academic Press, 2017), 407–15.
19. Pico Caroni, Flavio Donato, and Dominique Muller, "Structural Plasticity upon Learning: Regulation and Functions," *Nature Reviews Neuroscience* 13 (2012): 478–90.
20. S. Cavallaro et al., "Memory-Specific Temporal Profiles of Gene Expression in the Hippocampus," *Proceedings of the National Academy of Sciences of the United States of America*, 99 (2002): 16279–84; Y. Lin et al., "Activity-Dependent Regulation of Inhibitory Synapse Development by Npas4," *Nature* 455, no. 7217 (2008): 1198–204, https://doi.org/10.1038/nature07319. See also B. Hertler et al., "Temporal Course of Gene Expression during Motor Memory Formation in

Primary Motor Cortex of Rats," *Neurobiology of Learning and Memory* 136 (2016): 105–15.

21. Usain Bolt can run about 20 miles per hour. Bears have been clocked at 30 miles per hour. Here is a video of a bear chasing a deer though the woods: "Bear Hunting Deer," YouTube, December 15, 2012, https://www.youtube.com/watch?v=JqGiLMpZdBw&frags=pl%2Cwn. The deer is caught.
22. Suzana Herculano-Houzel, *The Human Advantage: How Our Brains Became Remarkable* (Cambridge, MA: MIT Press, 2016).
23. La Cerra and Bingham, *Origin of Minds*. Peggy La Cerra and Roger Bingham were the first biologists to teach me the widespread significance of energetic constraints in the mammalian brain.
24. Bill Schutt, *Cannibalism: A Perfectly Natural History* (Chapel Hill, NC: Algonquin Books, 2017). There is some evidence that under conditions of famine, for example, humans may eat placental material; see Jack Miles, *God: A Biography* (New York: Vintage Books, 1995). And a sixteenth-century medical text, *Compendium of materia medica*, makes reference to eating the placenta, perhaps after drying.
25. But see Corinne Purtill, "No Mothers in Human History Ate Their Own Placentas before the 1970s," Quartz, July 7, 2017, https://qz.com/1022404/no-mothers-in-human-history-ate-their-own-placentas-before-the-1970s.
26. YouTube has lots of videos of heroic mammalian mothers, such as this one: "Mother Squirrel Goes Nuts and Saves Baby!" YouTube, March 1, 2009, https://www.youtube.com/watch?v=T2wxVdo2osQ.
27. Sarah Blaffer Hrdy, *Mothers and Others: The Evolutionary Origins of Mutual Understanding* (Cambridge, MA: Belknap Press of Harvard University Press, 2009).
28. Ngogo Chimpanzee Project, accessed July 8, 2018, http://campuspress.yale.edu/ngogochimp/project.
29. A. Rusu, B. Knig, and S. Krackow, "Pre-reproductive Alliance Formation in Female Wild House Mice (*Mus domesticus*): The Effects of Familiarity and Age Disparity," *Acta Ethologica* 6, no. 2 (2004): 53–58.
30. K. Langergraber, J. Mitani, and L. Vigilant, "Kinship and Social Bonds in Female Chimpanzees (*Pan troglodytes*)," *American Journal of Primatology* 71 (2009): 840–51.
31. Adrian Viliami Bell and Katie Hinde, "Who Was Helping? The Scope for Female Cooperative Breeding in Early *Homo*," *PLoS One* 9, no. 3 (2013), https://doi.org/10.1371/journal.pone.0083667.
32. E. A. D. Hammock and L. J. Young, "Neuropeptide Systems and Social Behavior: Noncoding Repeats as a Genetic Mechanism for Rapid Evolution of Social Behavior," *Evolution of Nervous Systems* 3 (2017): 361–71.
33. A. Whiten, V. Horner, and F. B. M. de Waal, "Conformity to Cultural Norms of Tool Use in Chimpanzees," *Nature* 437, no. 7059 (2005): 737–40; C. P. Van Schaik, R. O. Deaner, and M. Y. Merrill, "The Conditions for Tool Use in Primates: Implications for the Evolution of Material Culture," *Journal of Human Evolution* 36, no. 6 (1999): 719–41.
34. Henry Gee insightfully reminds us of this fact in his book *The Accidental Species: Misunderstandings of Human Evolution* (Chicago: University of Chicago, 2013), 74.
35. Richard Wrangham, *Catching Fire: How Cooking Made Us Human* (New York: Basic Books, 2000).

36. Suzana Herculano-Houzel, "The Remarkable, yet Not Extraordinary, Human Brain as a Scaled-Up Primate Brain and Its Associated Cost," *Proceedings of the National Academy of Sciences of the United States of America* 109 (2012): 10661–68. See her TED talk here: Suzana Herculano-Houzel, "What Is So Special about the Human Brain?" (TED Talks), TEDGlobal, June 2013, https://www.ted.com/talks/suzana_herculano_houzel_what_is_so_special_about_the_human_brain.
37. Rachel N. Carmody, Gil S. Weintraub, and Richard W. Wrangham, "Energetic Consequences of Thermal and Nonthermal Food Processing," *Proceedings of the National Academy of Sciences of the United States of America* 108 (2011): 19199–203.
38. Marta Florio and Wieland B. Huttner, "Neural Progenitors, Neurogenesis and the Evolution of the Neocortex," *Development* 141 (2014): 2182–94.
39. This is from a favorite Flanders and Swann song: "The Hippopotamus Song (Mud, Mud, Glorious Mud)," YouTube, November 7, 2008, https://www.youtube.com/watch?v=1QW85kfakJc.

Chapter 2: Getting Attached

1. From Jeremy Holmes, *John Bowlby and Attachment Theory* (London: Routledge, 1993).
2. This prairie vole behavior was first noticed by Lowell Getz and Joyce Hofmann. L. L. Getz and J. E. Hofmann, "Social Organization in Free Living Prairie Voles, *Microtus ochrogaster*," *Behavioral Ecology and Sociobiology* 18 (1986): 275–82.
3. When I outlined this story to Stephen Colbert on *The Colbert Report*, he leaned in and said, "The prairie vole is a Christian"; "Patricia Churchland," *Colbert Report*, January 23, 2014, http://www.cc.com/video-clips/fykny6/the-colbert-report-patricia-churchland. See C. S. Carter et al., "Oxytocin, Vasopressin, and Sociality," *Progress in Brain Research* 170 (2008): 331–36; L. Young and B. Alexander, *The Chemistry between Us: Love, Sex and the Science of Attraction* (New York: Current Hardcover, 2012).
4. L. B. King et al., "Variation in the Oxytocin Receptor Gene Predicts Brain Region-Specific Expression and Social Attachment," *Biological Psychiatry* 80, no. 2 (2016): 160–69, https://doi.org/10.1016/j.biopsych.2015.12.008; E. A. D. Hammock and L. J. Young, "Neuropeptide Systems and Social Behavior: Noncoding Repeats as a Genetic Mechanism for Rapid Evolution of Social Behavior," *Evolution of Nervous Systems* 3 (2017): 361–71.
5. Here is Hume: "Reason, being cool and disengaged, is no motive to action, and directs only the impulse received from appetite or inclination, by showing us the means of attaining happiness or avoiding misery: Taste, as it gives pleasure or pain, and thereby constitutes happiness or misery, becomes a motive to action, and is the first spring or impulse to the desire and volition." David Hume, *A Treatise of Human Nature: A Critical Edition*, ed. David Fate Norton and Mary J. Norton (Oxford: Clarendon, 2007), book 4, pt. 1, sec. 3.
6. E. B. Keverne and K. M. Kendrick, "Oxytocin Facilitation of Maternal Behavior in Sheep," *Annals of the New York Academy of Sciences* 652 (1992): 83–101.
7. R. Corona and F. Levy, "Chemical Olfactory Signals and Parenthood in Mammals," *Hormones and Behavior* 68 (2015): 77–90.
8. C. Finkenwirth et al., "Strongly Bonded Family Members in Common Marmosets

Show Synchronized Fluctuations in Oxytocin," *Physiology and Behavior* 151 (2015): 246–51, https://doi.org/10.1016/j.physbeh.2015.07.034. Epub July 29, 2015.

9. J. P. Burkett et al., "Oxytocin-Dependent Consolation Behavior in Rodents," *Science* 351, no. 6271 (2016): 375–78.
10. Stephanie D. Preston and Frans De Waal, "Empathy: Its Ultimate and Proximate Bases," *Behavior and Brain Sciences* 25 (2002): 1–20.
11. Andrea E. Kudwa, Robert F. McGivern, and Robert J. Handa, "Estrogen Receptor β and Oxytocin Interact to Modulate Anxiety-like Behavior and Neuroendocrine Stress Reactivity in Adult Male and Female Rats," *Physiology and Behavior* 29 (2014): 287–96.
12. The medial preoptic area, or MPOA.
13. Z. Wu et al., "Galanin Neurons in Medial Preoptic Area Govern Parental Behavior," *Nature* 509 (2014): 325–30.
14. E. Scott et al., "An Oxytocin-Dependent Social Interaction between Larvae and Adult *C. elegans*," *Science Reports* 7, no. 1 (2017): 10122.
15. E. B. Keverne, "Mammalian Viviparity: A Complex Niche in the Evolution of Genomic Imprinting," *Heredity* 113 (2014): 138–44.
16. Johannes Kohl, Anita E. Autry, and Catherine Dulac, "The Neurobiology of Parenting: A Neural Circuit Perspective," *BioEssays* 39, no. 1 (2017): 1–11.
17. Stimulation can be done artificially, as it sometimes is in zoo mammals.
18. Jaak Panksepp, "Feeling the Pain of Social Loss," *Science* 302, no. 5643 (2003): 237–39; K. D. Broad, J. P. Curley, and E. B. Keverne, "Mother–Infant Bonding and the Evolution of Mammalian Social Relationships," *Philosophical Transactions of the Royal Society of London. Series B, Biological Sciences* 361, no. 1476 (2006): 2199–214.
19. A. Beyeler et al., "Organization of Valence-Encoding and Projection-Defined Neurons in Basolateral Amygdala," *Cell Reports* 22, no. 4 (2018): 905–18.
20. Kyle S. Smith et al., "Ventral Pallidum Roles in Reward and Motivation," *Behavior and Brain Research*, 196, no. 2 (2009): 155–67.
21. Such feelings depend not only on the neurochemical, but on the circuit where the receptors are located.
22. Don Wei et al., "Endocannabinoid Signaling in the Control of Social Behavior," *Trends in Neurosciences* 40 (2017): 385–96. See also Lin W. Hung et al., "Gating of Social Reward by Oxytocin in the Ventral Tegmental Area," *Science* 357 (2017): 1406–11.
23. Francesca R. D'Amato and Flaminia Pavone, "Modulation of Nociception by Social Factors in Rodents: Contribution of the Opioid System," *Psychopharmacology* 224 (2012): 189–200.
24. Well, things are a little more complicated, but that's the gist of it. See Michael Numan and Danielle S. Stolzenberg, "Medial Preoptic Area Interactions with Dopamine Neural Systems in the Control of the Onset and Maintenance of Maternal Behavior in Rats," *Frontiers in Neuroendocrinology* 30, no. 1 (January 2009): 46–64, https://doi.org/10.1016/j.yfrne.2008.10.002.
25. Michael Numan and Larry J. Young, "Neural Mechanisms of Mother–Infant Bonding and Pair Bonding: Similarities, Differences, and Broader Implications," *Hormones and Behavior* 77 (2016): 98–112.
26. A. M. Anacker et al., "Septal Oxytocin Administration Impairs Peer Affiliation via V1a Receptors in Female Meadow Voles," *Psychoneuroendocrinology* 68 (2016): 156–62, https://doi.org/10.1016/j.psyneuen.2016.02.025.

27. Anacker et al., "Septal Oxytocin Administration."
28. A. S. Smith and Z. Wang, "Hypothalamic Oxytocin Mediates Social Buffering of the Stress Response," *Biological Psychiatry* 76, no. 4 (2014): 281–88.
29. K. Gobrogge and Z. Wang, "Neuropeptidergic Regulation of Pair-Bonding and Stress Buffering: Lessons from Voles," *Hormones and Behavior* 76 (2015): 91–105.
30. There is evidence from chimpanzees in the wild that oxytocin is released during food sharing.
31. Matthew D. Lieberman, *Social: Why Our Brains Are Wired to Connect* (New York: Crown, 2013).
32. Alaska Wolves, "Wolf Pair Bonds," April 8, 2008, http://www.alaskawolves.org/Blog/CBB2EEB4-67FE-4796-B151-2A218C250613.html.
33. Sarah Blaffer Hrdy, *Mother Nature: Maternal Instincts and How They Shape the Human Species* (New York: Random House, 1999).
34. M. Kosfeld et al., "Oxytocin Increases Trust in Humans," *Nature* 435 (2005): 673–76.
35. P. S. Churchland and P. Winkielman. "Modulating Social Behavior with Oyxtocin: How Does It Work? What Does It Mean?" *Hormones and Behavior* 61 (2012): 392–99.
36. Hasse Walum, Irwin D. Waldman, and Larry J. Young, "Statistical and Methodological Considerations for the Interpretation of Intranasal Oxytocin Studies," *Biological Psychiatry* 79 (2016): 252.
37. Simon L. Evans et al., "Intranasal Oxytocin Effects on Social Cognition: A Critique," *Brain Research* 1580 (2014): 69–77.
38. D. A. Baribeau et al., "Oxytocin Receptor Polymorphisms Are Differentially Associated with Social Abilities across Neurodevelopmental Disorders," *Scientific Reports* 7 (2017): art. 11618, https://doi.org/10.1038/s41598-017-10821-0.
39. K. J. Parker et al., "Plasma Oxytocin Concentrations and OXTR Polymorphisms Predict Social Impairments in Children with and without Autism Spectrum Disorder," *Proceedings of the National Academy of Sciences of the United States of America* 111 (2014): 12258–63.
40. M. L. Boccia et al., "Immunohistochemical Localization of Oxytocin Receptors in Human Brain," *Neuroscience* 253 (2013): 155–64.
41. E. L. MacLean et al., "Effects of Affiliative Human-Animal Interaction on Dog Salivary and Plasma Oxytocin and Vasopressin," *Frontiers in Psychology* 8 (2017): 1606, https://doi.org/10.3389/fpsyg.2017.01606.

Chapter 3: Learning and Getting Along

1. M. Gervais and D. S. Wilson, "The Evolution and Functions of Laughter and Humor: A Synthetic Approach." *Quarterly Review of Biology* 80, no. 4 (2005): 395–430. Chimpanzees also laugh; for example, see M. Davila-Ross et al., "Aping Expressions? Chimpanzees Produce Distinct Laugh Types When Responding to Laughter of Others," *Emotion* 11, no. 5 (2011): 1013–20.
2. Amazingly, bees can learn to access a sweet reward by pulling a string, and bees that watch the successful learners learn faster than if they cannot watch. It appears that imitation of some kind is at work in the brains of these bees. Sylvain Alem et al., "Associative Mechanisms Allow for Social Learning and Cultural Transmission of String Pulling in an Insect," *PLoS Biology* 14, no. 12 (2016), https://doi.org/10.1371/journal.pbio.1002564.

3. Suzanne N. Haber, "Neuroanatomy of Reward: A View from the Ventral Striatum," in *Neurobiology of Sensation and Reward*, ed. Jay A. Gottfried (Boca Raton, FL: CRC Press, 2011), chap. 11, https://www.ncbi.nlm.nih.gov/books/NBK92777.
4. H. E. Atallah et al., "Neurons in the Ventral Striatum Exhibit Cell-Type-Specific Representations of Outcome during Learning," *Neuron* 82 (2014): 1145–56.
5. In the video "Grizzly Bear vs Caribou" (YouTube, October 4, 2006, https://www.youtube.com/watch?v=5SqqG_LUss0), notice that early on, the cubs lunge and retreat, exactly as the mom does. They are learning, partly by imitation. When the action heats up, the cubs watch from a distance. For cooperative hunting in wolves, see BBC Earth, *Frozen Planet*, "Pack of Wolves Hunt a Bison," YouTube, August 30, 2017, https://www.youtube.com/watch?v=8wl8ZxAaB2E.
6. W. Schultz, P. Apicella, and T. Ljungberg, "Responses of Monkey Dopamine Neurons to Reward and Conditioned Stimuli during Successive Steps of Learning a Delayed Response Task," *Journal of Neuroscience* 13, no. 3 (1993): 900–913.
7. I was fortunate to be in the lab at that time, collaborating with Sejnowski on our book, *The Computational Brain*.
8. Apologies to those who suppose that neurons can't talk.
9. R. S. Sutton and A. G. Barto, *Reinforcement Learning: An Introduction* (Cambridge, MA: MIT Press, 1998); R. S. Sutton and A. G. Barto, "Time-Derivative Models of Pavlovian Reinforcement," in *Learning and Computational Neuroscience: Foundations of Adaptive Networks*, ed. M. Gabriel and J. Moore (Cambridge, MA: MIT Press, 1990), 497–537.
10. P. R. Montague, P. Dayan, and T. J. Sejnowski, "A Framework for Mesencephalic Dopamine Systems Based on Predictive Hebbian Learning," *Journal of Neuroscience* 16, no. 5 (1996): 1936–47.
11. The classic citation for this work is Wolfram Schultz, Peter Dayan, and P. Read Montague, "A Neural Substrate of Prediction and Reward," *Science* 275, no. 5306 (1997): 1593–99.
12. Here is what the paper by Schultz, Apicella, and Ljungberg ("Responses of Monkey Dopamine Neurons") said: "None of the dopamine neurons showed sustained activity in the delay between the instruction and trigger stimuli that would resemble the activity of neurons in dopamine terminal areas, such as the striatum and frontal cortex. Thus, dopamine neurons respond phasically to alerting external stimuli with behavioral significance whose detection is crucial for learning and performing delayed response tasks. The lack of sustained activity suggests that dopamine neurons do not encode representational processes, such as working memory, expectation of external stimuli or reward, or preparation of movement. Rather, dopamine neurons are involved with transient changes of impulse activity in basic attentional and motivational processes underlying learning and cognitive behavior."
13. This description simplifies the anatomy. See Stephan Lammel, Byung Kook Lim, and Robert C. Malenka, "Reward and Aversion in a Heterogeneous Midbrain Dopamine System," *Neuropharmacology* 76 (2014): 351–59. I should also mention here that some neuroscientists prefer the somewhat more general anatomical name *striatum* to *nucleus accumbens*. The niceties of the differences between those two names do not matter for the general story that I am telling here.
14. There are also suggestions for the location of drug action. The relationship turns out to be rather more complex and confusing than initial results from rodent tests

indicated. See D. J. Nutt et al., "The Dopamine Theory of Addiction: 40 Years of Highs and Lows," *Nature Reviews Neuroscience* 16 (2015): 305–12.

15. See the commentary by Veronica A. Alvarez: "Clues on the Coding of Reward Cues by the Nucleus Accumbens," *Proceedings of the National Academy of Sciences of the United States of America* 113, no. 10 (2016): 2560–62.
16. Peter H. Rudebeck et al., "Prefrontal Mechanisms of Behavioral Flexibility, Emotion Regulation and Value Updating," *Nature Neuroscience* 16, no. 8 (2013): 1140–45.
17. Okihide Hikosaka, "The Habenula: From Stress Evasion to Value-Based Decision-Making," *Nature Reviews Neuroscience* 11 (2010): 503–13.
18. For a wonderful account of all this, see Terry Sejnowski, *The Deep Learning Revolution* (Cambridge, MA: MIT Press, 2018).
19. Jeremy Hsu, "Texas Hold'Em AI Bot Taps Deep Learning to Demolish Humans," IEEE Spectrum, March 2, 2017, https://spectrum.ieee.org/automaton/robotics/artificial-intelligence/texas-holdem-ai-bot-taps-deep-learning-to-demolish-humans.
20. T. W. Robbins and A. F. T. Arnsten, "The Neuropsychopharmacology of Fronto-executive Function: Monoaminergic Modulation," *Annual Review of Neuroscience* 32, no. 1 (2009): 267–87.
21. M. Konnikova, "The Struggles of a Psychologist Studying Self-Control," *New Yorker*, October 9, 2014, https://www.newyorker.com/science/maria-konnikova/struggles-psychologist-studying-self-control.
22. Silvia U. Maier, Aidan B. Makwana, and Todd A. Hare, "Acute Stress Impairs Self-Control in Goal-Directed Choice by Altering Multiple Functional Connections within the Brain's Decision Circuits," *Neuron* 18, no. 3 (2015): 621–31.
23. Kent C. Berridge, Terry E. Robinson, and J. Wayne Aldridge, "Dissecting Components of Reward: 'Liking,' 'Wanting,' and Learning," *Current Opinion in Pharmacology* 9, no. 1 (2009): 65–73.
24. A. D. Redish, "Addiction as a Computational Process Gone Awry," *Science* 306, no. 5703 (2004): 1944–47.
25. Terrence J. Sejnowski et al., "Prospective Optimization," *Proceedings of the IEEE* 102, no. 5 (2014): 799–811.
26. Terry Lohrenz et al., "Neural Signature of Fictive Learning Signals in a Sequential Investment Task," *Proceedings of the National Academy of Sciences of the United States of America* 104, no. 22 (2007): 9493–98.
27. More specifically, the ventral striatum. The nucleus accumbens is part of the ventral striatum.
28. Rosalyn Moran, quoted in Virginia Tech, "Keep Calm and Carry On: Scientists Make First Serotonin Measurements in Humans," Medical Xpress, April 30, 2018, https://medicalxpress.com/news/2018-04-calm-scientists-serotonin-humans.html.
29. Read Montague, quoted in Virginia Tech, "Keep Calm and Carry On."
30. A. P. Steiner and A. D. Redish, "The Road Not Taken: Neural Correlates of Decision Making in Orbitofrontal Cortex," *Frontiers in Neuroscience* 6 (2012): 131, https://doi.org/10.3389/fnins.2012.00131.
31. R. Eisenberger, "Achievement: The Importance of Industriousness," *Behavioral and Brain Sciences* 21 (1998): 412–13.
32. See Coursera, "Learning How to Learn: Powerful Mental Tools to Help You

Master Tough Subjects," accessed August 29, 2018, https://www.coursera.org/learn/learning-how-to-learn.

33. Ann M. Graybiel, "The Basal Ganglia and Cognitive Pattern Generators," *Schizophrenia Bulletin* 23, no. 3 (1997): 459–69.
34. Ann M. Graybiel, "Habits, Ritual, and the Evaluative Brain," *Annual Review of Neuroscience* 31 (2008): 359–87.

Chapter 4: Norms and Values

1. Alan Bennett, "Diary" [April 17, 2013], *London Review of Books* 36, no. 1 (January 9, 2014): 34–55.
2. For a review of children's learning through observation and helping, see Barbara Rogoff et al., "Firsthand Learning through Intent Participation," *Annual Review of Psychology* 54 (2003): 175–203. See also Ruth Paradise and Barbara Rogoff, "Side by Side: Learning by Observing and Pitching In," *Ethos* 37 (2009): 102–38.
3. Elizabeth A. Reynolds Losin et al., "Own-Gender Imitation Activates the Brain's Reward Circuitry," *Social Cognitive Affective Neuroscience* 7, no. 7 (2012): 804–10.
4. Susan Perry, "Social Traditions and Social Learning in Capuchin Monkeys (*Cebus*)," *Philosophical Transactions of the Royal Society of London. Series B, Biological Sciences* 366, no. 1567 (2011): 988–96.
5. See Knoxville Zoo, "This Is Einstein!" YouTube, July 26, 2008, https://www.youtube.com/watch?v=nbrTOcUnjNY.
6. V. Wörmann et al., "A Cross-Cultural Comparison of the Development of the Social Smile: A Longitudinal Study of Maternal and Infant Imitation in 6- and 12-Week-Old Infants," *Infant Behavioral Development* 35 (2012): 335–47, https://doi.org/10.1016/j.infbeh.2012.03.002. Epub June 19, 2012.
7. K. Tchalova and N. I. Eisenberger, "How the Brain Feels the Hurt of Heartbreak: Examining the Neurobiological Overlap between Social and Physical Pain," in *Brain Mapping: An Encyclopedic Reference*, ed. Arthur W. Toga (New York: Academic Press, 2015), 15–20; N. I. Eisenberger, "The Pain of Social Disconnection: Examining the Shared Neural Underpinnings of Physical and Social Pain," *Nature Reviews Neuroscience* 13 (2012): 421–34.
8. C. C. Ruff and E. Fehr, "The Neurobiology of Rewards and Values in Social Decision-Making," *Nature Reviews Neuroscience* 15 (2014): 549–62. See also M. J. Crockett et al., "Moral Transgressions Corrupt Neural Representations of Value," *Nature Neuroscience* 20, no. 6 (2017): 879–85.
9. P. La Cerra and R. Bingham, *The Origin of Minds* (New York: Harmony Books, 2002); J. Z. Siegel, M. J. Crockett, and R. J. Dolan, "Inferences about Moral Character Moderate the Impact of Consequences on Blame and Praise," *Cognition* 167 (2017): 201–11.
10. Roy F. Baumeister, *Evil: Inside Human Violence and Cruelty* (New York: Holt, 1997), 223.
11. M. V. Mestre et al., "Are Women More Empathetic than Men? A Longitudinal Study in Adolescence," *Spanish Journal of Psychology* 12, no. 1 (2009): 76–83; L. Christov-Moore et al., "Empathy: Gender Effects in Brain and Behavior," *Neuroscience & Biobehavioral Reviews* 4 (2014): 604–27.
12. V. Toccaceli et al., "Adult Empathy: Possible Gender Differences in Gene-Environment Architecture for Cognitive and Emotional Components in a Large

Italian Twin Sample," *Twin Research and Human Genetics* 21, no. 3 (2018): 214–26, https://doi.org/10.1017/thg.2018.19.

13. "Frans de Waal, Primatologist" [TED speaker, TEDx organizer], TED Talks, accessed August 29, 2018, https://www.ted.com/speakers/frans_de_waal.
14. Malini Suchak et al., "How Chimpanzees Cooperate in a Competitive World," *Proceedings of the National Academy of Sciences of the United States of America* 113, no. 36 (2016): 10215–20.
15. Alvin Roth et al., "Bargaining and Market Behavior in Jerusalem, Ljubljana, Pittsburgh and Tokyo: An Experimental Study," *American Economic Review* 81 (1991): 1068–95.
16. Joseph Henrich et al., "In Search of *Homo economicus*: Behavioral Experiments in 15 Small-Scale Societies," *American Economic Review* 91, no. 2 (2001): 73–78.
17. Ting Xiang, Terry Lohrenz, and P. Read Montague, "Computational Substrates of Norms and Their Violations during Social Exchange," *Journal of Neuroscience* 33, no. 3 (2013): 1099–108.

Chapter 5: I'm Just That Way

1. Bertrand Russell, *A History of Western Philosophy* (New York: Simon and Schuster, 1945), xxii.
2. Woo-Young Ahn et al., "Nonpolitical Images Evoke Neural Predictors of Political Ideology," *Current Biology* 24, no. 22 (2014): 2693–99.
3. Bear in mind that the scanner does not measure brain activity directly, but rather measures changes in oxygen levels in the blood. The blood oxygen level–dependent (BOLD) measure is generally accepted as an indirect signal of neuron activity.
4. The Wilson-Patterson inventory presents a list of topics, and subjects can select "Yes," "?," or "No," which are scored as 3, 2, and 1, respectively. See T. J. Bouchard Jr. et al., "Evidence for the Construct Validity and Heritability of the Wilson-Patterson Conservatism Scale: A Reared-Apart Twins Study of Social Attitudes," *Personality and Individual Differences* 34 (2003): 959–69.
5. Ahn et al., "Nonpolitical Images Evoke Neural Predictors," 2693.
6. M. D. Dodd et al., "The Political Left Rolls with the Good and the Political Right Confronts the Bad: Connecting Physiology and Cognition to Preferences," *Philosophical Transactions of the Royal Society of London. Series B, Biological Sciences* 367, no. 1589 (2012): 640–49, https://doi.org/10.1098/rstb.2011.0268.
7. J. R. Alford, C. L. Funk, and J. R. Hibbing, "Are Political Orientations Genetically Transmitted?" *American Political Science Review* 99, no. 2 (2005): 153–67.
8. T. J. Bouchard Jr. et al., "Sources of Human Psychological Differences: The Minnesota Study of Twins Reared Apart," *Science* 250, no. 4978 (1990): 223–28.
9. Jonathan Flint, Ralph J. Greenspan, and Kenneth S. Kendler, *How Genes Influence Behavior* (New York: Oxford University Press, 2010), 25.
10. For the data on twins reared apart, see Bouchard et al., "Evidence for the Construct Validity."
11. Dodd et al., "Political Left Rolls with the Good."
12. Gordon D. A. Brown, Corey L. Fincher, and Lukasz Walasek, "Personality, Parasites, Political Attitudes, and Cooperation: A Model of How Infection Prevalence Influences Openness and Social Group Formation," *Topics in Cognitive Science* 8 (2016): 98–117.
13. See also this comprehensive and fair-minded article: John R. Hibbing, Kevin

B. Smith, and John R. Alford, "Differences in Negativity Bias Underlie Variations in Political Ideology," *Behavioral and Brain Sciences* 37 (2014): 297–350. The authors explore the ways in which psychological and physiological responses to negative features of the environment differ along ideological lines. Open peer commentary follows; it is useful to see how other scientists respond to the results and their interpretation and, in turn, how Hibbing, Smith, and Alford address criticism.

14. L. Pessoa, "On the Relationship between Emotion and Cognition," *Nature Reviews Neuroscience* 9 (2008): 148–58.

Chapter 6: Conscience and Its Anomalies

1. "Life of Bishop Hacket," in *The Literary Remains of Samuel Taylor Coleridge*, vol. 3, ed. Henry Nelson Coleridge (London: W. Pickering, 1838), 186.
2. Patricia Smith Churchland, *Neurophilosophy: Towards a Unified Understanding of the Mind/Brain* (Cambridge, MA: MIT Press, 1986).
3. Robert D. Hare, *Without Conscience: The Disturbing World of the Psychopaths among Us* (New York: Guilford, 1993). Kent Kiehl's book *The Psychopath Whisperer: The Science of Those without a Conscience* (New York: Crown, 2014) is a brilliant successor to Hare's book. See also *Encyclopedia of Mental Disorders*, s.v. "Hare Psychology Checklist," accessed July 9, 2018, http://www.minddisorders.com/Flu-Inv/Hare-Psychopathy-Checklist.html#ixzz4wcrb2Ifk.
4. Hare wisely drew from the earlier work of Hervey M. Cleckley, an American psychiatrist who was a keen observer of patients in a locked neuropsychiatric hospital. Cleckley identified sixteen central features that he thought defined psychopaths. Hervey M. Cleckley, *The Mask of Sanity; an Attempt to Reinterpret the So-Called Psychopathic Personality* (Oxford: Mosby, 1941) (and many editions thereafter). Hare's unique contribution was to carefully evaluate each item in Cleckley's list, to add criteria, and to assemble a diagnostic tool—the "Checklist for Psychopathy"—that could be widely adopted to enable consistency across discussions and reports concerning psychopathy. This effort was highly successful in fostering research from diverse labs and regions so that progress could be made in understanding the brain basis of the disorder.
5. See Kiehl, *Psychopath Whisperer.* In chapter 3, Kiehl explains that using the checklist appropriately requires training, and he briefly illustrates that training by considering each trait in the test and determining whether a well-known historical figure has that trait in some degree.
6. S. D. Hart, R. D. Hare, and T. J. Harpur, "The Psychopathy Checklist-Revised (PCL-R): An Overview for Researchers and Clinicians," in *Advances in Psychological Assessment*, vol. 8, ed. J. C. Rosen and P. McReynolds (New York: Plenum, 1992), 103–30; R. D. Hare, *The Hare Psychopathy Checklist-Revised* (Toronto: Multi-Health Systems, 1991). The revised checklist is a little shorter than the original. Copyright limitations prevent me from listing the traits.
7. For example, J. C. Prichard, *A Treatise on Insanity and Other Disorders Affecting the Mind* (London: Sherwood, Gilbert and Piper, 1835).
8. Kiehl, *Psychopath Whisperer.*
9. Kiehl, *Psychopath Whisperer*, chap. 7.
10. C. A. Ficks, L. Dong, and I. D. Waldman, "Sex Differences in the Etiology of Psy-

chopathic Traits in Youth," *Journal of Abnormal Psychology* 123, no. 2 (2014): 406–11, https://doi.org/10.1037/a0036457; J. M. Horan et al., "Assessing Invariance across Sex and Race/Ethnicity in Measures of Youth Psychopathic Characteristics," *Psychological Assessment* 27, no. 2 (2015): 657–68, https://doi.org/10.1037/pas0000043.

11. C. J. Patrick, M. M. Bradley, and P. J. Lang, "Emotion in the Criminal Psychopath: Startle Reflex Modulation," *Journal of Abnormal Psychology* 102, no. 1 (1993): 82–92.
12. Kiehl, *Psychopath Whisperer*, chap. 5.
13. Ana Seara-Cardoso et al., "Anticipation of Guilt for Everyday Moral Transgressions: The Role of the Anterior Insula and the Influence of Interpersonal Psychopathic Traits," *Scientific Reports* 6 (2016): art. 36273, https://doi.org/10.1038/srep36273.
14. K. J. Yoder, E. C. Porges, and J. Decety, "Amygdala Subnuclei Connectivity in Response to Violence Reveals Unique Influences of Individual Differences in Psychopathic Traits in a Nonforensic Sample," *Human Brain Mapping* 36, no. 4 (2015): 1417–28.
15. J. Decety et al., "Socioemotional Processing of Morally-Laden Behavior and Their Consequences on Others in Criminal Psychopaths," *Human Brain Mapping* 36, no. 6 (2015): 2015–26.
16. Nathaniel E. Anderson and Kent Kiehl also emphasize this point, in "Psychopathy and Aggression: When Paralimbic Dysfunction Leads to Violence," *Current Topics in Behavioral Neuroscience* 17 (2014): 369–93. This genetic disorder is known as Urbach-Wiethe disease.
17. Anderson and Kiehl (in "Psychopathy and Aggression") also emphasize this point.
18. For a particularly clear and readable account of the scope and limits of fMRI, see R. A. Poldrack, *The New Mind Readers: The Power, Limits and Future of Brain Imaging* (Princeton, NJ: Princeton University Press, 2018).
19. M. R. Dadds et al., "Polymorphisms in the Oxytocin Receptor Gene Are Associated with the Development of Psychopathy," *Development and Psychopathology* 26, no. 1 (2014): 21–31.
20. E. Viding and E. J. McCrory, "Genetic and Neurocognitive Contributions to the Development of Psychopathy," *Development and Psychopathology* 24 (2012): 969–83.
21. G. Kochanska and S. Kim, "Toward a New Understanding of Legacy of Early Attachments for Future Antisocial Trajectories: Evidence from Two Longitudinal Studies," *Development and Psychopathology* 24 (2012): 783–806.
22. Martin H. Teicher et al., "The Effects of Childhood Maltreatment on Brain Structure, Function and Connectivity," *Nature Reviews Neuroscience* 17 (2016): 652–66.
23. M. T. Teicher and J. A. Samson, "Annual Research Review: Enduring Neurobiological Effects of Childhood Abuse and Neglect," *Journal of Child Psychology and Psychiatry* 57, no. 3 (2016): 241–66, https://doi.org/10.1111/jcpp.12507.
24. A. L. Schaller, S. A. Lakhani, and B. S. Hsu, "Pediatric Traumatic Brain Injury," *South Dakota Medicine* 68, no. 10 (2015): 457–61.
25. Larissa MacFarquhar, *Strangers Drowning: Impossible Idealism, Drastic Choices, and the Urge to Help* (New York: Penguin, 2015). There is certainly a distinction

between individuals who have an unusually strong urge to help, and those who suffer from scrupulosity, though biology being what it is, sometimes that distinction is rather blurry.

26. David Greenberg and Jonathan D. Huppert, "Scrupulosity: A Unique Subtype of Obsessive-Compulsive Disorder," *Current Psychiatry Reports* 12 (2010): 282–89.
27. M. Inozu, A. N. Karanci, and D. A. Clark, "Why Are Religious Individuals More Obsessional? The Role of Mental Control Beliefs and Guilt in Muslims and Christians," *Journal of Behavior Therapy and Experimental Psychiatry* 43, no. 3 (2012): 959–66.
28. Alphonsus de Liguori, *Selected Writings*, ed. Frederick M. Jones, The Classics of Western Spirituality (New York: Paulist Press, 1999), 322.
29. G. E. Ganss, ed., *Ignatius of Loyola: The Spiritual Exercises and Selected Works* (New York: Paulist Press, 1991), 77–78.
30. This is a useful website for information on scrupulosity: OCD Center of Los Angeles, "Moral Scrupulosity in OCD: Cognitive Distortions," June 17, 2014, http://ocdla.com/moral-scrupulosity-ocd-cognitive-distortions-3405.
31. Fletcher Wortmann, *Triggered: A Memoir of Obsessive-Compulsive Disorder* (New York: St. Martin's Press, 2012).
32. For a case report of OCD starvation, see Dinesh Dutt Sharma, Ramesh Kumar, and Ravi Chand Sharma, "Starvation in Obsessive-Compulsive Disorder Due to Scrupulosity," *Indian Journal Psychiatry* 48, no. 4 (2006): 265–66, https://doi.org/10.4103/0019-5545.31563.
33. I should perhaps add that we were graduate students and had never heard of such a phenomenon. Not understanding that he was suffering a serious mental disorder, we fully expected he would come to his senses when he became really hungry. Then, we thought, surely he would quit the foolishness. Now, of course, I would try to find his family and alert them, or at least find a doctor to intervene. All too often, ignorance is not bliss.
34. "The Penn Inventory of Scrupulosity (PIOS)," accessed July 9, 2018, http://www.psytoolkit.org/survey-library/scrupulosity-pios.html. Other inventories designed to assay OCD include these: the Maudsley Obsessive-Compulsive Inventory, the Perfectionism Cognitions Inventory, the Multidimensional Anger Inventory, and the Barron Ego Strength Scale.
35. A. M. Graybiel and S. L. Rauch, "Toward a Neurobiology of Obsessive–Compulsive Disorder," *Neuron* 28 (2000): 343–47; Claire M. Gillan and Trevor W. Robbins, "Goal-Directed Learning and Obsessive–Compulsive Disorder," *Philosophical Transactions of the Royal Society of London. Series B, Biological Sciences* 369, no. 1655 (2014): 20130475, https://doi.org/10.1098/rstb.2013.0475.
36. Dr. Seuss, *Oh, the Places You'll Go!* (New York: Random House, 1990).
37. See *Made of Metaphors* (blog), "Balancing Act," accessed July 9, 2018, http://madeofmetaphors.com/balancing-act.
38. Adam Smith, *The Theory of Moral Sentiments*, ed. D. D. Raphael and A. L. Macfie (Oxford: Clarendon, 1976), pt. 3, chap. 1, sec. 2. First published 1749.
39. Plato, *The First Alcibiades: A Dialogue on the Nature of Man*, trans. Floyer Sydenham and Thomas Taylor (Jon W. Fergus, 2016), 132C–33C. First published 1804. See also Seneca, *Letters from a Stoic*, trans. Richard M. Gummere (Digireads.com, 2017), 11.8–10, 25.5–6.

Chapter 7: What's Love Got to Do with It?

1. James Q. Wilson, *The Moral Sense* (New York: Free Press, 1993), 96.
2. See Ruth Paradise and Barbara Rogoff, "Side by Side: Learning by Observing and Pitching In," *Ethos* 37, no. 1 (2009): 102–38.
3. This was essentially the heart of the criticism by Thomas Hurka, a mainstream moral philosopher at the University of Toronto, in his discussion of my work at the Chicago meeting of the American Philosophical Association in 2012.
4. Among the notable exceptions are Julian Baggini, Simon Blackburn, Ronnie deSousa, David Edmonds, Owen Flanagan, Mark Johnson, Walter Sinott-Amstrong, and Nigel Warburton. One and all, they have given me courage.
5. Simon Blackburn's version is short and captures the essence of Socrates's ideas: Simon Blackburn, *Being Good: A Short Introduction to Ethics* (New York: Oxford University Press, 2001). For a lengthier discussion that is both brilliant and personal, see Mark Johnson, *Morality for Humans: Ethical Understanding from the Perspective of Cognitive Science* (Chicago: Chicago University Press, 2014). For a witty discussion with "Mr. Deity" on this subject, see "Mister Deity and the Philosopher," YouTube, August 24, 2011, https://www.youtube.com/watch?v=pwf6QD-REMY.
6. Thomas Nagel, "Ethics without Biology," in *Mortal Questions* (Cambridge: Cambridge University Press, 1979), 142–46.
7. *Stanford Encyclopedia of Philosophy Archive*, s.v. "Morality and Evolutionary Biology," by William FitzPatrick, accessed April 20, 2018, https://plato.stanford.edu/archives/spr2016/entries/morality-biology.
8. See also Christine Korsgaard, *The Sources of Normativity* (New York: Cambridge University Press, 1996).
9. Immanuel Kant, *The Groundwork to the Metaphysics of Morals*, trans. H. J. Paton (New York: Harper and Row, 1964), 57. First published 1785.
10. In accordance with official Nazi policy, about eight hundred children who would now be described as having autism spectrum disorder were killed at the Am Spiegelgrund clinic in Vienna. It has been documented that Hans Asperger was complicit in this horror. See Edith Sheffer, *Asperger's Children: The Origins of Autism in Nazi Vienna* (New York: Norton, 2018).
11. See also Jesse R. Prinz, *The Emotional Construction of Morals* (New York: Oxford University Press, 2007), 48, for a view that decries the arrogance of supposing there is a faculty of pure reason that taps into the universal moral laws.
12. George Bernard Shaw, *Maxims for Revolutionists* (San Bernardino, CA: Dossier Press, 2016), 1.
13. Simon Blackburn, *Ethics: A Very Short Introduction* (New York: Oxford University Press, 2001).
14. See, for example, Bernard Williams, *Ethics and the Limits of Philosophy* (San Bernardino, CA: Fontana Press, 1985).
15. Joshua Greene calls our tendency of preferring to care for our own children over those many in some other part of the planet a "species typical moral limitation." Greene adopts the utilitarian principle of impartiality, which says that we ought not favor family over unknown others. For an insightful review, see Thomas Nagel, "You Can't Learn about Morality from Brain Scans," *New Republic*, November 1, 2013, https://newrepublic.com/article/115279/joshua-greenes-moral-tribes-reviewed-thomas-nagel.

16. See also G. Kahane et al., "'Utilitarian' Judgments in Sacrificial Moral Dilemmas Do Not Reflect Impartial Concern for the Greater Good," *Cognition* 134 (2015): 193–209.
17. Owen J. Flanagan, *Varieties of Moral Personality* (Cambridge, MA: Harvard University Press, 1991), loc. 757, Kindle.
18. Blackburn, *Being Good*, loc. 851, Kindle.
19. This is how Ilya Farber characterizes much of the history of moral theories. Conversation with author, June 2017.
20. J. P. Sheppard, D. Raposo, and A. K. Churchland, "Dynamic Weighting of Multisensory Stimuli Shapes Decision-Making in Rats and Humans," *Journal of Vision* 13, no. 6 (2013), https://doi.org/10.1167/13.6.4; A. L. Juavinett, J. C. Erlich, and A. K. Churchland, "Decision-Making Behaviors: Weighing Ethology, Complexity and Sensorimotor Compatibility," *Current Opinion in Neurobiology* 49 (2018): 42–50.
21. P. Grimaldi, H. Lau, and M. A. Basso, "There Are Things That We Know That We Know, and There Are Things That We Do Not Know That We Do Not Know: Confidence in Decision-Making," *Neuroscience and Biobehavioral Reviews* 55 (2015): 88–97.
22. B. W. Brunton, M. M. Botvinick, and C. D. Brody, "Rats and Humans Can Optimally Accumulate Evidence for Decision-Making," *Science* 340 (2013): 95–98.
23. Sheppard et al., "Dynamic Weighting."
24. Cendri A. Hutcherson, Benjamin Bushong, and Antonio Rangel, "A Neurocomputational Model of Altruistic Choice and Its Implications," *Neuron* 87 (2015): 451–62; Nathaniel D. Daw, Yael Niv, and Peter Dayan, "Uncertainty-Based Competition between Prefrontal and Dorsolateral Striatal Systems for Behavioral Control," *Nature Neuroscience* 8 (2005): 1704–11.
25. Camillo Padoa-Schioppa and Katherine E. Conen, "Orbitofrontal Cortex: A Neural Circuit for Economic Decisions," *Neuron* 96, no. 4 (2017): 736–54.
26. Z. Jonke, S. Habenschuss, and W. Maass, "Solving Constrain Satisfaction with Networks of Spiking Neurons," *Frontiers in Neuroscience*, March 30, 2016, https://doi.org/10.3389/fnins.2016.00118.
27. For an experimental inquiry into the nature of social interactions, see Andreas Hula, P. Read Montague, and Peter Dayan, "Monte Carlo Planning Method Estimates Planning Horizons during Interactive Social Exchange," *PLoS Computational Biology* 11, no. 6 (2015): e1004254.
28. Frans de Waal, *The Age of Empathy: Nature's Lessons for a Kinder Society* (New York: Harmony Books, 2009). The first book of De Waal's that I read—*Good Natured* (Cambridge, MA: Harvard University Press, 1996)—greatly helped me to regard morality not as something otherworldly and Platonic, but as part of our nature.
29. Jane Goodall, *Through a Window: My Thirty Years with the Chimpanzees of Gombe* (New York: Houghton, Mifflin, Harcourt, 2010).
30. Cristina Marquez et al., "Prosocial Choice in Rats Depends on Food-Seeking Behavior Displayed by Recipients," *Current Biology* 25 (2015): 1736–45.
31. See also the amazing video of an experiment showing one rat rescuing another, from research by Inbar Ben-Ami Bartal in Peggy Mason's lab at the University of Chicago: "Biological Roots of Empathically Motivated Helping Behaviour, Strong Evidence," YouTube, December 9, 2011, https://www.youtube.com/watch?v=3jkOwYKBJEI.

32. Liran Samuni et al., "Oxytocin Reactivity during Intergroup Conflict in Wild Chimpanzees," *Proceedings of the National Academy of Sciences of the United States of America* 114, no. 2 (2017): 268–73.
33. Simon Gachter and Jonathan F. Schultz, "Intrinsic Honesty and the Prevalence of Rule Violations across Societies," *Nature* 531 (2016): 496–99. See also Maria Konnikova, "How Norms Change," *New Yorker*, October 11, 2017, https://www.newyorker.com/science/maria-konnikova/how-norms-change?.
34. *Morality for Humans* is the title of Mark Johnson's wonderful book.
35. Alison Gopnik, *The Philosophical Baby: What Children's Minds Tell Us about Truth, Love and the Meaning of Life* (New York: Farrar, Straus and Giroux, 2009); Michael Tomasello, *A Natural History of Human Morality* (Cambridge MA: Harvard University Press, 2016).
36. Wilson, *Moral Sense*.
37. Franz Boas, *The Central Eskimo*, Sixth Annual Report of the Bureau of Ethnology to the Secretary of the Smithsonian Institution, 1884–1885 (Washington, DC: Government Printing Office, 1888).
38. The songs in these duels apparently could get very elaborate, typical songs were all about the opponent and his flaws, and the outcome was determined by how much the audience enjoyed the songs performed. A song duel may not resolve a dispute, whereupon a highly regulated physical exchange may take place. A widening of the dispute to involve other members is strongly discouraged. For an example of a song duel, see "Song Duel," YouTube, November 14, 2009, https://www.youtube.com/watch?v=nuoy4dPbaP4&frags=pl%2Cwn. See also Penelope Eckett and Russell Newmark, "Central Eskimo Song Duels: A Contextual Analysis of Ritual Ambiguity," *Ethnology*, 19, no. 2 (1980): 191–211.
39. Many thanks to Kevin Mitchell for this point, which he raised in conversation with me at the Cold Spring Harbor Laboratory meeting "Wiring the Brain," March 24–28, 2015, but also in his review of my earlier book: Patricia S. Churchland, *Braintrust: What Neuroscience Tells Us about Morality* (Princeton NJ: University Press, 2011). See his review here: Kevin Mitchell, "Where Do Morals Come From?" *Wiring the Brain* (blog), June 13, 2011, http://www.wiringthebrain.com/2011/06/where-do-morals-come-from.html.
40. Ralph Adolphs and David Anderson, *The Neuroscience of Emotion in Humans and Animals* (Princeton, NJ: Princeton University Press, 2018); Lisa Feldman Barrett, *How Emotions Are Made: The Secret Life of the Brain* (New York: Houghton, Mifflin, Harcourt, 2017).
41. Elephants have about 257 billion neurons, which seems astonishing, but about 97% of the neurons (251 billion) are in neither the cortex nor subcortical structures, but in the cerebellum. It is still a mystery why elephants have such a remarkably large cerebellum.
42. Suzana Herculano-Houzel, "The Human Brain in Numbers: A Linearly Scaled-Up Primate Brain," *Frontiers in Human Neuroscience*, November 9, 2009.
43. E. A. D. Hammock and L. J. Young, "Neuropeptide Systems and Social Behavior: Noncoding Repeats as a Genetic Mechanism for Rapid Evolution of Social Behavior," *Evolution of Nervous Systems* 3 (2017): 361–71.
44. William Kaufman, ed., *Meditations*, Dover Thrift Editions (Mineola, NY: Dover, 1996). Different classical scholars translate the passage in somewhat different ways. This translation seems especially clear.

Chapter 8: The Practical Side

1. Richard Sorabji, *Moral Conscience through the Ages: Fifth Century B.C. to the Present* (Chicago: University of Chicago Press, 2014); Paul Strohm, *Conscience: A Very Short Introduction* (Oxford: Oxford University Press, 2011).
2. Confucius commented in the same vein: Real knowledge is to know the extent of one's ignorance.
3. Some historians portray this aspect as the core of what Socrates meant by "know thyself," but my reading suggests that for Socrates, self-knowledge of one's traits is only part of the far-ranging attitude of striving to be realistic in general, despite temptations to make things easy for ourselves.
4. See my discussion in Patricia S. Churchland, "Free Will, Habits, and Self-Control," in *Touching a Nerve: The Self as Brain* (New York: Norton, 2013), chap. 7.
5. For a very thoughtful discussion of why psychopaths should not be held responsible for their actions, see Paul Litton, "Criminal Responsibility and Psychopathy: Do Psychopaths Have a Right to Excuse?" in *Handbook on Psychopathy and Law*, ed. Kent A. Kiehl and Walter Sinnott-Armstrong (Oxford: Oxford University Press, 2013), 275–96. And for an equally thoughtful but different opinion, see Samuel H. Pillsbury, "Why Psychopaths Are Responsible," in *Handbook on Psychopathy and the Law*, 297–318.
6. See, for example, Stephen J. Morse, "Preventative Detention of Psychopaths and Offenders," in *Handbook on Psychopathy and the Law*, 321–45.
7. I am tempted to self-plagiarize here from my book *Touching a Nerve*. In Chapter 7, "Free Will, Habits, and Self-Control," I go into this matter in much more detail than I can here.
8. E. Nahmias et al., "Surveying Freedom: Folk Intuitions about Free Will and Moral Responsibility," *Philosophical Psychology* 18, no. 5 (2005): 561–84.
9. The movie telling this remarkable story, *The Wizard of Lies*, is based on the book by Diana B. Henriques: *The Wizard of Lies: Bernie Madoff and the Death of Trust* (New York: Times Books/Holt, 2011).
10. This explanation is highly simplified. See A. Davenport, *Basic Criminal Law: The Constitution, Procedure, and Crimes*, 5th ed. (New York: Pearson, 2017).
11. My words, from *Touching a Nerve*, 180.
12. Stephen J. Morse, "Lost in Translation? An Essay on Law and Neuroscience," *Current Legal Issues* 13 (2010): 533.
13. See especially Owen D. Jones, Jeffrey D. Schall, and Francis X. Shen, *Law and Neuroscience* (New York: Kluwer, 2014); and Kiehl and Sinott-Armstrong, *Handbook on Psychopathy and Law.*
14. For example, see Bonnie Badenoch, *The Heart of Trauma* (New York: Norton, 2018).
15. Sherry Turkle, *Alone Together: Why We Expect More of Technology and Less of Each Other* (New York: Basic Books, 2011).
16. Center for Practical Wisdom, "About," accessed September 12, 2018, http://wisdomresearch.org/Arete/About.aspx.
17. C. E. Barrett, S. E. Arambula, and L. J. Young, "The Oxytocin System Promotes Resilience to the Effects of Neonatal Isolation on Adult Social Attachment in Female Prairie Voles," *Translational Psychiatry* 5 (2015): e606, https://doi.org/10.1038/tp.2015.73.
18. See, for example, Steven Brill, *Tailspin: The People and Forces behind America's*

Fifty-Year Fall and Those Fighting to Reverse It (New York: Knopf, 2018); Thomas Frank, *Rendezvous with Oblivion: Reports from a Sinking Society* (New York: Holt, 2018).

19. Steve Schmidt, in Hayley Miller, "GOP Strategist Quits 'Corrupt' Part of 'Feckless Cowards,' Will Vote for Democrats," *HuffPost*, June 20, 2018, https://www.huffingtonpost.com/entry/steve-schmidt-renounces-gop_us_5b2a4ce8e4b05d6c16c96a05; Richard Parker, "American Internment Camps," *New York Times*, June 20, 2018, https://www.nytimes.com/2018/06/20/opinion/american-internment-camps.html.
20. CNN, "Sessions Admits Policy Is a Deterrent," June 19, 2018, https://www.cnn.com/videos/politics/2018/06/19/sessions-defends-controversial-immigration-policy-deterrent-sot.cnn/video/playlists/senator-jeff-sessions.
21. Aleksandr I. Solzhenitsyn, *The Gulag Archipelago 1918–1956*, trans. Thomas P. Whitney and Harry Willetts (New York: HarperPerennial, 2007), 174.

Index

Note: Illustrations are indicated with *italic* page numbers. Endnotes are indicated by n after the page number.